8

Springer Series in Chemical Physics

Edited by J. P. Toennies, V. I. Goldanskii

Springer Series in Chemical Physics
Editors: V. I. Goldanskii R. Gomer F. P. Schäfer J. P. Toennies

Volume 1 **Atomic Spectra and Radiative Transitions**
By I. I. Sobelman

Volume 2 **Surface Crystallography by LEED** Theory, Computation and Structural Results By M. A. Van Hove, S. Y. Tong

Volume 3 **Advances in Laser Chemistry** Editor: A. H. Zewail

Volume 4 **Picosecond Phenomena**
Editors: C. V. Shank, E. P. Ippen, S. L. Shapiro

Volume 5 **Laser Spectroscopy** Basis Concepts and Instrumentation
By W. Demtröder

Volume 6 **Laser-Induced Processes in Molecules** Physics and Chemistry
Editors: K. L. Kompa, S. D. Smith

Volume 7 **Excitation of Atoms and Broadening of Spectral Lines**
By I. I. Sobelman, L. A. Vainshtein, E. A. Yukov

Volume 8 **Spin Exchange** Principles and Applications in Chemistry and Biology
By Yu. N. Molin, K. M. Salikhov, K. I. Zamaraev

Volume 9 **Secondary Ion Mass Spectrometry SIMS II**
Editors: A. Benninghoven, C. A. Evans, Jr., R. A. Powell, R. Shimizu, H. A. Storms

Volume 10 **Lasers and Chemical Change**
By A. Ben-Shaul, Y. Haas, K. L. Kompa, R. D. Levine

Volume 11 **Liquid Crystals of One-and Two-Dimensional Order**
Editors: W. Helfrich, G. Heppke

Volume 12 **Gasdynamic Lasers** By S. A. Losev

Volume 13 **Atomic Many-Body Theory**
By I. Lindgren, J. Morrison

Volume 14 **Picosecond Phenomena II**
Editors: R. Hochstrasser, W. Kaiser, C. V. Shank

Volume 15 **Vibrational Spectroscopy of Adsorbates**
Editor: R. F. Willis

Yu. N. Molin K. M. Salikhov K. I. Zamaraev

Spin Exchange

Principles and Applications in Chemistry and Biology

With 68 Figures

Springer-Verlag Berlin Heidelberg New York 1980

Professor Yuri N. Molin
Professor Kev M. Salikhov
Institute of Chemical Kinetics and Combustion, Academy of Sciences of the USSR
Siberian Branch, Novosibirsk 630090, USSR

Professor Kirill I. Zamaraev
Institute of Catalysis, Academy of Sciences of the USSR
Siberian Branch, Novosibirsk 630090, USSR

Series Editors

Professor Vitalii I. Goldanskii
Institute of Chemical Physics
Academy of Sciences
Vorobyevskoye Chaussee 2-b
Moscow V-334, USSR

Professor Robert Gomer
The James Franck Institute
The University of Chicago
5640 Ellis Avenue
Chicago, IL 60637, USA

Professor Dr. Fritz Peter Schäfer
Max-Planck-Institut für
Biophysikalische Chemie
D-3400 Göttingen-Nikolausberg
Fed. Rep. of Germany

Professor Dr. J. Peter Toennies
Max-Planck-Institut für Strömungsforschung
Böttingerstraße 6–8
D-3400 Göttingen
Fed. Rep. of Germany

ISBN-13: 978-3-642-67668-0 e-ISBN-13: 978-3-642-67666-6
DOI: 10.1007/978-3-642-67666-6

Library on Congress Cataloging in Publication Data
Molin, Yuri Nikolaevich. Spin exchange. (Springer series in chemical physics ; v. 8) Revised translation of Spinovyĭ obmen. Bibliography: p. Includes index. 1. Electron paramagnetic resonance spectroscopy. 2. Spin exchange. 3. Collision spectroscopy. 4. Solution (Chemistry) I. Salikhov, Kev Minullinovich, joint author. II. Zamaraev, Kirill Il'ich, joint author. III. Zamaraev, Kirill Il'ich. Spinovyĭ obmen. IV. Title. V. Series. QD96.E4M64 543'.0877 80-16693

Softcover reprint of the hardcover 1st edition 1980

Offset printing: Beltz Offsetdruck, Hemsbach/Bergstr. Bookbinding: J. Schäffer oHG, Grünstadt.
2153/3130-543210

Preface

Spin exchange between paramagnetic particles is one of the few elementary bimolecular processes in solutions that is readily accessible to detailed experimental studies and to strict theoretical description. Due to this fact spin exchange is now widely used in solving various problems of chemistry, physics, and molecular biology. Spin exchange allows one to study in detail elementary acts of collisions of molecules. By use of spin exchange one can obtain quantitative information on collision rates of molecules in solutions, steric hinderance in collisions, electrostatic interactions in collisions of charged particles, and overlap of electron orbitals in collisions. Particularly valuable is the possibility of using spin exchange to study collisions in such complex systems as polymer solutions, multicomponent mixtures, heterogeneous mixtures, and especially biological systems.

This monograph contains a comprehensive description of spin exchange in solutions. It consists of four chapters. The introductory chapter contains a brief summary of the physical principles of spin exchange and outlines the range of its main applications. In this chapter a historic background of research in this field is also given. The second chapter contains a comprehensive presentation of the theory of spin exchange including that of its spectroscopic manifestations. The third chapter is devoted to a detailed description of the technique of experimental studies of spin exchange. The last – the fourth – chapter summarizes the results of experimental studies of spin exchange and considers various applications of spin exchange in chemistry, molecular physics, and biology.

This volume is a revised edition of the monograph "Spinowii obmjen. Teorija i fisiko-chimitscheskije priloschenija" (Spin Exchange. Theory and Physicochemical Applications) published in Russian in 1977. In comparison with the Russian edition this volume contains a more detailed description of biological applications of spin exchange as well as various techniques of studying it. At the same time a number of sections that were in the Russian edition have been omitted here. Among the latter are those about spin exchange between atoms and free radicals in gases, between triplet excitons in crystals, as well as about the role of spin exchange in quenching of positronium.

The present book is intended for scientists working in the fields of chemical kinetics, molecular physics, molecular biology, and chemical magnetic radiospectroscopy. The volume can also serve as a textbook for students in studying the applications of magnetic radiospectroscopy in chemistry and biology.

Our interest in studying spin exchange was to a great extent inspired by our teacher Prof. V.V. Voyevodsky. It was under the influence of his ideas on the role of weak interactions in chemistry that the main trends of our research on the applications of spin exchange were defined.

We are grateful to A.B. Doktorov, A.I. Kokorin, A.T. Nikitaev, V.N. Parmon, and G.I. Skubnevskaya with whose collaboration many results reflected in the book were obtained. We are also much indebted to V.M. Berdnikov for his useful remarks and discussions, and to L.M. Weiner for his assistance in writing the section on biological applications of spin exchange. We are grateful to all the authors and publishers of those journals who kindly permitted us to reproduce their figures in the book. We also express our thanks to Drs. Hyde and Atkins for their valuable advice concerning the structure of the book which has been taken into account in preparing this edition.

Novosibirsk, July 1980

Yu.N. Molin
K.M. Salikhov
K.I. Zamaraev

Contents

1. Introduction

Spin exchange is a change of spin states of paramagnetic particles in collisions which is induced by exchange interaction of partners that occurs when their electron orbitals overlap. Exchange interaction arises as a result of the dependence of the Coulomb repulsion of electrons on the orientation of their spins (see, e. g. [1.1,2]). According to the Pauli exclusion principle the spatial distributions of electrons are different depending on whether their spins are parallel or antiparallel. Two electrons with antiparallel spins can be simultaneously found at the same point of space, while two electrons with parallel spins can not. As a result the Coulomb repulsion is stronger for the first case. When the energy of exchange interaction is small enough, exchange interaction between two particles A and B can be described by a well-known Hamiltonian [1.1,2]

$$\hat{V} = \hbar J_{AB}(r)\hat{S}_A \hat{S}_B \tag{1.1}$$

where $J_{AB}(r)$ is the so-called exchange integral, $\hat{S}_A$ and $\hat{S}_B$ are spin operators of interacting particles A and B, r is the distance between the particles, $\hbar$ is the Planck constant divided by 2 π. For the simplest case $S_A = S_B = 1/2$ Hamiltonian (1.1) has two eigenstates corresponding to the total spin value $S = 0$ (the singlet state) and $S = 1$ (the triplet state), the energy splitting between them being $\hbar\, J_{AB}(r)$.

The spin dependence of the electron interaction energy becomes important only when their orbitals overlap. Since the overlap of the electron wave functions decreases rapidly with the distance r between the interacting atoms, the energy of exchange interaction also decreases rapidly with the increase of r.

At short distances, such as those between atoms in molecules, when the overlap is substantial and the energy of exchange interaction is large enough as compared to the energy of thermal motion, this interaction results in the formation of chemical bonds between atoms. In solids such interactions, when their energy exceeds the energy of thermal motion, result in the appearance of ferromagnetic or antiferromagnetic properties.

Spin exchange in solutions is induced by far weaker exchange interactions that occur when the particles approach to a Van der Waals distance. This energy is much smaller than the energy of chemical bonds. In this situation exchange interaction is quite well described by Hamiltonian (1.1). This simplifies essentially the theoretical treatment of spin exchange.

Moreover, at Van der Waals distances J_{AB} is usually smaller than the energy of thermal motion. Thus spin exchange practically does not perturb the character of the thermal motion of interacting molecules. This circumstance furthermore simplifies the theoretical description of spin exchange kinetics.

Though exchange interactions between paramagnetic particles during their collisions are very small, they can nevertheless essentially change the states of the spins of these particles. Usually these changes consist in an exchange of spin states between the partners, i. e., a change in the state of electron spin of one of the particles is fully compensated by an opposite change in the electron spin state of its partner. In this most common situation spin exchange can be represented by a scheme shown in Fig.1.1.

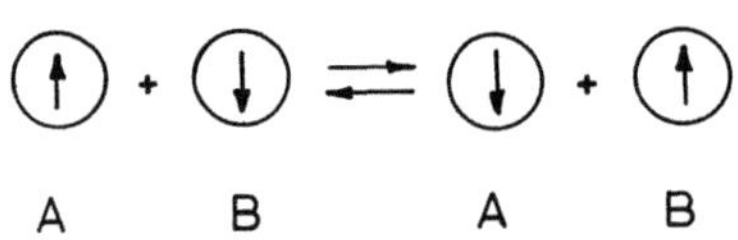

Fig.1.1. Schematic representation of a spin exchange process between two paramagnetic particles. The vertical arrows indicate the spin orientations and their exchange as a result of a collision

In dilute solutions spin exchange takes place via binary collisions of paramagnetic particles:

$$A(\uparrow) + B(\downarrow) \rightleftarrows A(\downarrow) + B(\uparrow) \tag{1.2}$$

and by analogy with bimolecular chemical reactions its rate can be represented as

$$W = K_e C_A C_B \tag{1.3}$$

where C_A and C_B are the concentrations of particles A and B, and K_e is the rate constant.

Most simply the rate constants of spin exchange can be measured from the ESR spectra of interacting particles. (For introduction to ESR spectroscopy see [1.3,4]). Depending on specific conditions the ESR lines can be broadened, shifted, or narrowed by spin exchange. Reliable methods have been developed that allow one to measure the rate constants of spin exchange in all those situations. From ESR spectra one can measure even extremely large rate constants of spin exchange (up to 10^{11} l/mol · s), but the smallest values of rate constants that can be measured by this method usually do not exceed 10^7 l/mol · s.

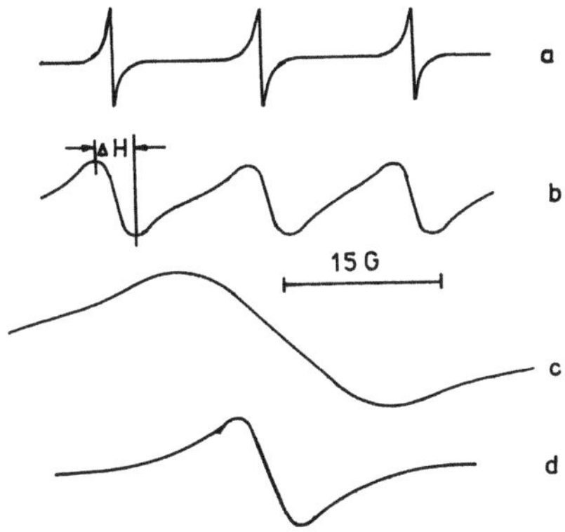

Fig.1.2a-d. The influence of spin exchange on the ESR spectrum of a nitroxide free radical. The rate of spin exchange increases from the top to the bottom spectrum as a result of increase of the radical concentration: (a) no spin exchange; (b) exchange broadening; (c) collapse of the hyperfine structure; (d) exchange narrowing

Figure 1.2 represents a typical example of the change of the ESR spectra as a result of spin exchange. In this figure the ESR spectra of a stable nitroxide free radical are given for different concentrations of this compound in solution. Free radicals of this type are often used in studies of spin exchange. The spectrum (a) corresponds to very dilute solutions of the free radical, when the influence of spin exchange on the spectrum can be neglected. The spectrum consists of three lines due to hyperfine interaction of the unpaired electron with the nuclear magnetic moment of the nitrogen atom. The increase in the rate of spin exchange at larger free radical concentrations results in broadening of the ESR lines (see spectrum b) and in a small shift of the outer lines towards the center of the spectrum.

Further increase in free radical concentration leads to the collapse of the three hyperfine lines into a broad single line (spectrum c) and then to the narrowing of this single line (spectrum d).

The rate constant K_e can be most easily found from the dependence of the line broadening on the concentration of paramagnetic particles. This broadening occurs as a result of shortening of the lifetime of the electron spin states due to spin exchange in agreement with the uncertainty principle. The magnitude of this broadening can be related to the rate constant of spin exchange K_e and the concentration of paramagnetic particle, C, as

$$A \cdot (\Delta H - \Delta H_0) = K_e \cdot C \qquad (1.4)$$

where ΔH_0 and ΔH are the line widths in the absence and in the presence of spin exchange. The coefficient A can be easily calculated for any particular pair of interacting particles. (See Chap. 3 for a detailed discussion). For the specific example of spin exchange between nitroxide radicals

$$A = \frac{3}{2}\, 1.52 \cdot 10^7\ G^{-1}s^{-1} \quad .$$

Thus K_e can be calculated in this case from the concentration dependence of the ESR line broadening with the help of the relation

$$K_e = \frac{3}{2}\, 1.52 \cdot 10^7\ (\Delta H - \Delta H_0)\ /C \qquad (1.5)$$

where ΔH and ΔH_0 are measured between the points of the maximum slope (see Fig.1,2b).

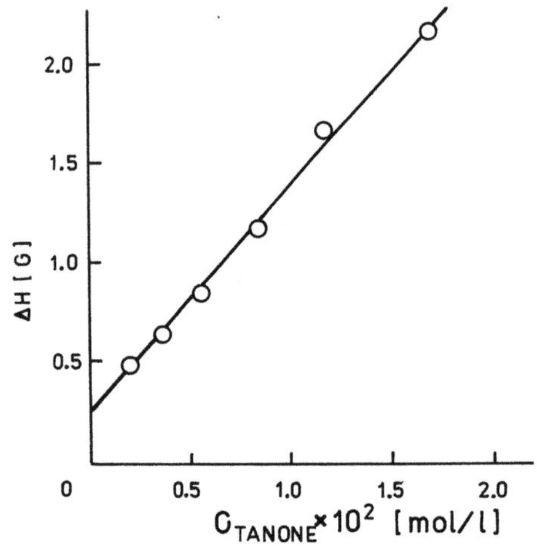

Fig.1.3. The concentration dependence of the ESR linewidth for the nitroxide free radical

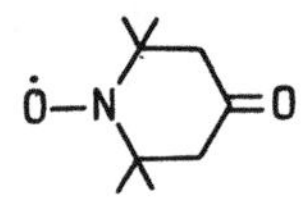

in aqueous solution. The concentration broadening is a result of spin exchange between the radical particles

As an example, Fig.1.3 presents the experimentally observed concentration dependence of the linewidth of the ESR spectrum central component for the nitroxide free radical

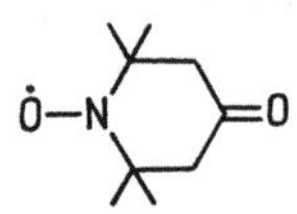 .

It is seen that in accordance with (1.4) the line width is indeed a linear function of the radical concentration. The rate constant of spin exchange, determined from the slope of the straight line in Fig.1.3 is equal to $2.7 \cdot 10^9$ l/mol · s.

Spin exchange rate constants can also be measured with the help of electron double resonance, electron spin echo, nuclear magnetic resonance and other methods. These methods, however, are more complicated and are not yet so widespread in spin exchange study as the ESR technique.

As is usually the case in chemical kinetics, the spin exchange rate constant can be presented as the product of the diffusion rate constant K_D of binary collisions by the average efficiency p of one collision

$$K_e = p\, K_D \quad . \tag{1.6}$$

For spherical particles the collision efficiency depends on the average energy J of exchange interaction during a collision, on the duration τ_c of collisions as well as on the spin values of interacting particles. The larger J and τ_c are, the larger is the efficiency p of collisions in spin exchange. Since the energy of exchange interaction reflects the degree of overlapping of unpaired electron orbitals, the efficiency p is larger, the stronger and the longer these orbitals overlap in the course of collisions. In the extreme case when the energy of exchange interaction during a collision and the duration of the collision are large enough (the case of strong exchange) the limiting stage of spin exchange is diffusion. In the other extreme situation—at very small values of exchange interaction energy and of collision duration (the case of weak exchange)—the efficiency becomes far smaller than unity and the process is limited by spin exchange itself. These extreme situations are similar to the well-known extreme situations in liquid-phase chemical reactions when, depending on the ratio between the duration of a collision and the characteristic time of the reaction, the limiting stages of the process are either the diffusion or the reaction itself [1.5,6].

The formulae expressing p through J and τ_c for various spin exchange processes can be easily obtained with the help of the theory presented in Chap. 2.

In the simplest case when the electron spins of both the colliding particles $S_A = S_B = 1/2$ the efficiency of collisions can be expressed as

$$p = \frac{1}{2}\,\frac{J^2\tau_c^2}{1 + J^2\tau_c^2} \quad . \tag{1.7}$$

One can see from (1.7) that in case of strong exchange ($J^2\tau_c^2 >> 1$) the efficiency of collisions does not depend on the energy of exchange interaction and on the duration of collisions τ_c and becomes equal to its maximum possible value $p = 1/2$. In the other extreme case of weak exchange ($J^2\tau_c^2 << 1$) the efficiency p is given by

$$p = \frac{1}{2} J^2\tau_c^2 \tag{1.8}$$

and decreases with the decrease of both J and τ_c.

For nonspherical particles the act of spin exchange is more complicated since the energy of exchange interaction between particles A and B is dependent on their mutual orientation which during a collision changes due to random thermal motion. In this case the overlap of orbitals of unpaired electrons is sufficiently large only for some definite mutual orientations of colliding particles A and B. To take this fact into account one should additionally introduce into (1.3) for the exchange rate constant a steric factor $f < 1$

$$K_e = p\, f\, K_D \quad . \tag{1.9}$$

For charged particles the value of the spin exchange rate constant must also be sensitive to the electrostatic interactions during collisions. These interactions can influence both the diffusion constant K_D and the collision efficiency p (due to changes in the duration of collisions).

Thus the spin exchange rate in solutions is influenced by several factors. Note that the same factors usually influence also the rate of chemical reactions. Spin exchange can be used to study the role of each of these factors. In this sense spin exchange can be considered as a process modeling certain features of chemical reactions.

In order to obtain information about the mechanism of collisions in solutions from spin exchange data it is necessary to develop special experimental techniques for separation of the influence of various factors on K_e. As is discussed in detail in Chaps.3,4, this can be done by looking at the trends in variation of the K_e values with the variation of solvent viscosity, temperature, or of the structure of interacting particles. As can be seen, for example, from (1.6,7) the trends in the variation of the rate constant of spin exchange with the variation of viscosity should be of the opposite character in the extreme cases of strong and weak exchange. For strong exchange $K_e = K_D/2$ and thus K_e should increase with the decrease in viscosity, since $K_D \sim \eta^{-1}$. For weak exchange, when p is given by (1.8),

$$K_e = \frac{1}{2} J^2 \tau_c^2 \cdot K_D \qquad (1.10)$$

and K_e is expected to increase with viscosity since $\tau_c \sim \eta$.

Note that the role played by the above-mentioned various factors affecting chemical reactions in liquids can be found from spin exchange studies far more easily than from studying the kinetics of the chemical reactions themselves.

Table 1.1. Examples demonstrating the influence of various factors on the rate constants of spin exchange between particle A and B. (See Chap. 4 for references and detailed discussion)

No	Particle A	Particle B	Solvent	K_e [l/mol · s]
1	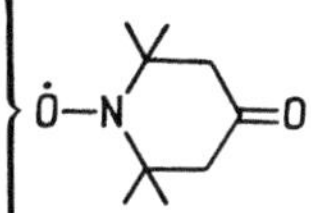	$Ni(en)_2(H_2O)_2^{2+}$	water	$2.1 \cdot 10^9$
2			30% water + 70% glycerol	$2.3 \cdot 10^8$
3		Heminpyridinate	pyridine	$2.4 \cdot 10^9$
4		Ferrihemoglobin	water	$< 10^8$
5		Oxidized cytochrome C	water	$< 10^8$
6	$CH_3\dot{C}HOH$	$Co(H_2O)_6^{2+}$	water	$2.6 \cdot 10^9$
7	$(CH_3)_2\dot{C}OH$		water	$2.0 \cdot 10^9$
8	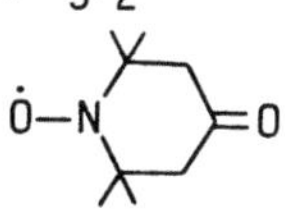		water	$0.9 \cdot 10^9$
9	$(SO_3)_2NO^{2-}$	$Cr(H_2O)_6^{3+}$	water	$1.5 \cdot 10^{10}$
10		$Cr(en)_2(SCN)^{2+}$	water	$8.8 \cdot 10^9$
11		$Cr(en)(C_2O_4)_2^-$	water	$2.3 \cdot 10^9$
12		$Cr(CN)_6^{3-}$	water	$1.7 \cdot 10^8$
13	$Cr(diphenyl)_2J$	$Co(C_3H_7OH)_2Cl_2$	propanol	$2.0 \cdot 10^9$
14		$Co(C_3H_7OH)_m(ClO_4)_n^{2-n}$	propanol	$3.0 \cdot 10^7$

Table 1.1 demonstrates the influence of various factors on the values of rate constants of spin exchange. The difference in the values of K_e for processes 1 and 2 characterizes quantitatively variation in the rate of collisions as solvent viscosity is changed. A dramatic drop in the rate constant

for spin exchange of a stable nitroxide free radical with Fe(III) as one goes from simple heminpyridinate complex (process 3) to active sites of ferrihemoglobin and cytochrome C (processes 4 and 5) reflects steric hindrances for collisions of the radical with Fe(III) in protein molecules. Comparison of K_e for processes 6,7 and 8 demonstrates that the rates of spin exchange for active (alcohol) and stable (nitroxide) free radicals are of the same order of magnitude. The variation of K_e in the sequence of processes 9 to 12 characterizes quantitatively the influence of the electrostatic interaction on the rate of collisions between electrically charged particles. The large difference between the rates of spin exchange for processes 13 and 14 is connected with the difference in the degree of the unpaired electron delocalization on the outer, available for collisions, atoms of Co(II) complexes. In the $Co(C_3H_7OH)_2Cl_2$ complex chlorine atoms with large unpaired electron densities are available for collisions with the Cr $(diphenyl)_2J$ complex, while in the $Co(C_3H_7OH)_m(ClO_4)_n^{2-n}$ complex the unpaired electron density is small for all the outer atoms.

From the above considerations on the character of the information contained in the values of spin exchange rate constants it is possible to outline the range of applications of spin exchange.

First of all one can use spin exchange to directly measure the rate constant of diffusion collisions as well as to study the role of steric hindrance during collisions in solutions. It should be especially noted that spin exchange makes it possible to study directly how collisions are affected by microscopic properties of liquids. These properties which are so far very difficult for theoretical analysis become especially important at short distances between interacting particles. The microscopic properties of the medium can be particularly essential for collisions of charged and other strongly solvated particles, molecules of complicated geometric shape such as, e.g., macromolecules including biopolymers. Of great interest is the possibility to directly measure the number of collisions in heterogeneous and, in particular, microheterogeneous systems including membranes and other biological systems. Very promising for the applications in molecular biology is, e.g., the fact that spin exchange can be used to study the accessibility for collisions of paramagnetic centers in enzymes or of spin labels selectively attached to biopolymers.

Information on exchange interactions that can be extracted in the situation of weak exchange interaction is also of interest in several aspects. For example, one can follow the dependence of the spin interaction efficiency on the structure of colliding particles and, in particular, to study how this interaction is conducted through the chain of chemical and intermolecular bonds separating the paramagnetic centers of colliding particles. These investigations are of great interest for chemical kinetics, since exchange interactions determining the spin exchange efficiency must influence in a similar way other more complex chemical and physical processes accompanied by the rearrangement of spin degrees of freedom. Examples of such processes are the ortho-para-conversion of positronium and the quenching of the triplet state via the exchange mechanism. Exchange interactions seem also to essentially affect the recombination of radicals by changing the rate of the singlet-triplet transitions in radical pairs. The data on the exchange interactions are also useful for understanding the mechanisms of other processes involved by weak overlap of electron orbitals, e.g., of the outer-sphere electron transfer reactions.

Among the pioneering works that provided the basis for systematic studies of spin exchange the following should be mentioned. Soon after the discovery of the ESR by ZAVOISKY [1.7] it was found that the shape of ESR spectra is very sensitive to exchange interactions. In 1947 GORTER and VAN VLECK [1.8] noticed that the ESR lines of solid paramagnetics observed by Zavoisky had a far smaller width as compared to what was expected from dipole-dipole interactions. They explained this effect as a result of averaging the dipole-dipole broadening by exchange interaction. In 1948 VAN VLECK [1.9] developed a detailed theory of this phenomenon. In the same year BAGGULEY and GRIFFITHS [1.10] discovered a merging of the ESR lines of two nonequivalent paramagnetic center ions in a crystal as a result of exchange interaction between them. Simultaneously PRYCE [1.11] gave a theoretical explanation to this effect.

The first observations of the influence of exchange interactions on the ESR spectra of paramagnetic particles in solution were made in 1952 by GARSTENS and LIEBSON [1.12] when they studied water solutions of the Mn(II) salts. They discovered narrowing of ESR lines in concentrated solutions induced by exchange interaction.

An important role in understanding the spectral manifestations of spin exchange was played by the work of ANDERSON and WEISS [1.13,14] in which they analyzed the influence of an accidental change in the Larmor frequency of spins on the shape of the ESR spectra. Though these authors did not study the case of spin exchange in solutions specially, the general model they proposed allowed one to explain from one point of view such different manifestations of spin exchange as broadening, shift, merging, and narrowing of ESR lines.

A study of spin exchange between atoms in gases was started by WITTKE and DICKE in 1956 [1.15]. They were the first to calculate from ESR spectra a cross section of spin exchange between hydrogen atoms.

Systematic research of spin exchange in solutions was started by a pioneering work of PAKE and TUTTLE [1.16]. They were the first to express in a clear form the idea that, as distinguished from crystals, exchange interactions in diluted solutions are provided by binary collisions of paramagnetic particles and to introduce the concept of collision efficiency in spin exchange. They also came to the very important conclusion that the principal mechanism of the concentration broadening of ESR lines in diluted solutions of free radicals is exchange, rather than dipole-dipole interaction. This offered a possibility of experimental research of spin exchange not only from the narrowing of ESR lines usually observed only in concentrated solutions, but also from the line broadening in diluted solutions.

The dependence of collision efficiency in spin exchange in solutions on the energy of exchange interaction J and on the duration τ_c of a collision was first analyzed theoretically by KIVELSON [1.17]. The theoretical investigations in this direction were widely developed later.

A further step was the study of intramolecular spin exchange in biradicals which provided a sort of a bridge between the extreme cases of static exchange interactions in crystals and of spin exchange via collisions in solutions. In the studies on spin exchange involving complexes of transition metals it was established how this process is affected by the presence of several unpaired electrons as well as by the rapid relaxation of the electron spin [1.18]. As a result of these studies the range of chemical and physical applications of spin exchange became considerably broader.

An important role in promoting the studies on the application of spin exchange in chemistry, physics, and biology was played by works on stable nitroxide radicals. Wide use of various radicals of this type for chemical research was provided in the first place by the works of Rozantsev, Rassat, and Buchachenko and for biological research by the works of McConnell and Likhtenshtein (for corresponding references see [1.19-21]).

At the present time the theory describing the spectroscopic manifestations of spin exchange seems to be essentially completed though some important details are still to be further developed and refined. Therefore in recent years the research of exchange interactions in collisions has been mainly focussed on elucidating the details of the elementary act of spin exchange and on its various applications.

2. Theory of Spin Exchange

The theory of bimolecular spin exchange process deals with the three main problems: calculation and estimation of an exchange integral for various pairs of paramagnetic particles, calculation of spin exchange rate constants, and analysis of experimental manifestations of spin exchange. Problems connected with exchange interaction and its experimental manifestations have been discussed in numerous monographs on quantum chemistry and on electron spin resonance (ESR). The present chapter, therefore, does not dwell on these problems in detail, but summarizes only the most essential results. As to the theory of spin exchange rate constants, it is systematically and comprehensively set forth for the first time in this monograph. We tried to collect and generalize here the results of all available theoretical calculations of rate constants of spin exchange between various paramagnetic particles in diluted solutions.

2.1 Exchange Interaction Between Two Paramagnetic Particles

Spin exchange is induced by exchange interaction between colliding paramagnetic particles. Therefore, before considering the kinetics of spin exchange processes it seems reasonable to discuss the physical nature of exchange interaction, its connection with the electron structure of paramagnetic particles, and some ways of estimating the strength of exchange energies for systems that are really delt with in experiments on spin exchange.

Dependence of Interaction Energy on Spin

The energy of electron states is known to be determined by the configuration of the nuclei and the spatial distribution of the electrons in a system. But owing to the Pauli exclusion principle the spatial distribution of electrons is dependent on their spin alignment. This entails the spin dependence of the electron energy. Indeed, according to Pauli two electrons with parallel spin alignment cannot be simultaneously at the same point of space. However, if the electron spins are antiparallel there is no such restriction and their Coulomb repulsion is stronger. The spin state of electrons in a system influences both interaction between them and their interaction with the nuclei.

The spin dependence of the interaction energy of paramagnetic particles appears only when their electron orbitals overlap. Spin exchange in solutions occurs when the paramagnetic particles A and B approach each other to a distance approximately equal to the sum of their Van der Waals radii, so the overlap of unpaired electrons is comparatively small. Therefore the spins of paramagnetic particles S_A and S_B are sufficiently good quantum numbers, and the spin fo the whole system A+B represents the vector sum of the moments $\underline{S}_A$ and $\underline{S}_B$. In this situation the energy of interaction between the molecules A and B will be small as compared with that of their intramolecular electron excitation. In this case the interaction will split the ground electron level of the system into 1+2S sublevels (where S is the smaller of the two spin quantum numbers S_A and S_B) with different values of the total spin of the system equal to $S_A + S_B$, $S_A + S_B - 1, \ldots, |S_A - S_B|$. By way of illustration Fig. 2.1 is a graph of the energy level splitting depending on the distances between A and B. At large distances between A and B when their electron orbitals do not overlap the energy levels with different values of the total spin coincide. However, when the particles approach each other and their orbitals overlap, these levels split. The diagram of the A and B states (see Fig.2.1) becomes particularly simple when at least one of the paramagnetic particles, say B, has spin 1/2. Then the level of the pair A+B splits into only two sublevles corresponding to the total spin $S_1 = S_A + 1/2$ and $S_2 = S_A - 1/2$. For example, the interaction of two free radicals ($S_A = 1/2$; $S_B = 1/2$) yields a triplet state ($S_1 = 1$) and a singlet state ($S_2 = 0$).

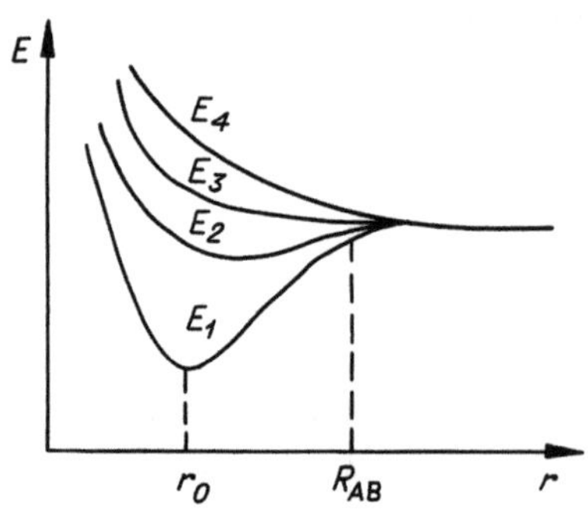

Fig. 2.1. Multiplet splitting of states when orbitals of A and B overlap. The splitting depends on the distance between the particles since the overlap of A and B orbitals is a function of this distance. The states with the energies E_1, E_2, etc., correspond to different values of the total spin of the interacting particles. r_0 is the equilibrium interatomic distance in the molecule AB, R_{AB} is the sum of the Van der Waals radii of particles A and B

Spin-Hamiltonian of Exchange Interaction

The quantitative analysis of spin exchange between paramagnetic particles when these collide is essentially facilitated by the fact that the states of weakly interacting particles can be represented as eigenstates of an equivalent spin-Hamiltonian of exchange interaction [2.1-5][1]

$$\hat{V} = \hbar J_{AB}(r)\hat{S}_A\hat{S}_B \quad . \tag{2.1}$$

Indeed, the eigenstates of this Hamiltonian are multiplets with the total spins $S_A + S_B$, $S_A + S_B - 1,\ldots, |S_A - S_B|$ so that (2.1) reflects correctly the multiplicity of the states of the pair A+B presented in Fig. 2.1. Hamiltonian (2.1) contains a single parameter J_{AB} called an exchange integral which characterize the value of eigenstate splitting. If the spin of at least one of the interacting partners, say S_B, is equal to 1/2 and there are only two states with energies E_1 and E_2 corresponding to the total spin S_1 and S_2 then one parameter J_{AB} is sufficient for the effective Hamiltonian (2.1) to give correctly the value of multiplet splitting as well. To accomplish this the exchange integral must be assumed to equal

$$\hbar J_{AB} = (E_1 - E_2) / (S_A + 1/2) \quad . \tag{2.2}$$

When the particles interact with S_A, $S_B > 1/2$ the number of eigenvalues for (2.1) is more than two (see Fig.2.1) and in the general case one parameter J_{AB} is not enough to describe the mutual position of all multiplets. In the case under consideration, however, when the interaction between A and B is relatively weak the situation is essentially simplified and the spin-Hamiltonian (2.1) describes equally well the spin dependence of the interaction energy of the two paramagnetic particles [2.3,7].

The Nature of Exchange Interaction. The Two-Electron System

It has been noted above that exchange interaction has an electrostatic nature and the multiplet splitting of the coupled A+B states reflects the dependence of the spatial distribution of electrons on the value of the total

---

1 In the theory of exchange interaction a spin-Hamiltonian is written in the form $V = -\hbar J(1/2 + 2\,\hat{S}_A\hat{S}_B)$ [2.6]. We make use of form (2.1) because it is in this form of the spin-Hamiltonian that the rates of spin exchange are calculated in the original works on the spin exchange theory. This allows us to maintain the uniformity with the original results. It is evident that this choice of form for the spin-Hamiltonian is by no means a matter of principle. A good correspondence is established between the two spin-Hamiltonians of exchange interaction by the equation $J_{AB} = -2J$.

$\hbar = h/2\pi$ (normalized Planck's constant)

spin of electrons in the system. By way of illustration let us analyze the exchange interaction of two atoms each of which has one electron.

Let a and b be orbitals of atoms A and B occupied by electrons. At great distances r between A and B when the interatomic interaction is still relatively small but the electron clouds of A and B are already overlapping it is a good approximation to represent the wave function of A and B as the product of orbitals a and b (valence bond approximation[2] [2.7, 10]).

In this case the wave functions of the singlet and triplet states of the system A+B are equal [2.11] to

$$^1\Psi_{AB} = [a(1)b(2) + a(2)b(1)]\ (2 + 2|S_{ab}|^2)^{-\frac{1}{2}}\ ,$$
$$^3\Psi_{AB} = [a(1)b(2) - a(2)b(1)]\ (2 - 2|S_{ab}|^2)^{-\frac{1}{2}}\ . \qquad (2.3)$$

where 1 and 2 refer to electrons. The quantity S_{ab} is the overlap integral of the orbitals a and b and is equal to

$$S_{ab} = \int a^* b\ dv \equiv \langle a|b\rangle \quad .$$

Note that two orbitals are orthogonal if $S_{ab} = 0$.

According to (2.2) for the system under consideration the exchange integral is equal to the difference of energies of the states with the wave functions $^3\Psi_{AB}$ and $^1\Psi_{AB}$. The electron Hamiltonian of the systems is equal [2.11] to

$$\hat{H}_e = \hat{H}(1) + \hat{H}(2) + e^2/r_{12} \qquad (2.4)$$

where $\hat{H}_{(k)} = \hat{T}_{(k)} + V_{(k)}$ is the sum of the kinetic (T) and the potential (V) energy of the kth electron; V represents the Coulomb interaction of the electron with the nuclei of both atoms; $V_{(k)} = V_{(k)A} + V_{(k)B}$. The last term in (2.4) characterizes the mutual repulsion of electrons. Calculating the electron energy $E = \int \Psi^* \hat{H}_e \Psi dv$ with the help of wave functions (2.3), for the energy difference of the triplet and the singlet we obtain the following expression [2.10]:

$$\begin{aligned} J_{AB} &= \hbar^{-1}(^3E_{AB} - {^1E_{AB}}) \\ &= -2\hbar^{-1}\ [S^*_{ab} \langle a|V_A|b\rangle + S_{ab} \langle b|V_B|a\rangle - |S_{ab}|^2(\langle a|V_B|a\rangle \\ &\quad + \langle b|V_A|b\rangle + \langle ab|e^2/r_{12}|ab\rangle) + \langle ab|e^2/r_{12}|ba\rangle]\ /\ (1 - |S_{ab}|^4)\ . \end{aligned} \qquad (2.5)$$

---

2 At very great distances between atoms the Heitler-London approximation does not describe correctly the dependence of the exchange integral on distance [2.8]. However, at medium distances (on the order of Van der Waals radii) the results obtained by virtue of the asymptotic theory [2.8] and in the framework of the valence bond approximation practically coincide [2.9].

The physical nature of singlet-triplet splitting is revealed with particular distinctness if using the wave functions (2.3) we calculate the electron distribution in the singlet ${}^1\Psi_{AB}$ and the triplet ${}^3\Psi_{AB}$ state. To do this we introduce one- and two-electron densities, $P_1(\underline{r})$ and $P_2(\underline{r}_1,\underline{r}_2)$.

The pair distribution function for one electron to be in a volume $dv(\underline{r}_1)$ provided the second electron is in $dv(\underline{r}_2)$ is equal to $|\Psi_{AB}(\underline{r}_1,\underline{r}_2)|^2$. Therefore, for the singlet and the triplet states we obtain, respectively,

$$
\begin{aligned}
{}^1P_2(\underline{r}_1,\underline{r}_2) &= [q(\underline{r}_1,\underline{r}_2) + q_e(\underline{r}_1,\underline{r}_2)]/(1+|S_{ab}|^2) \ , \\
{}^3P_2(\underline{r}_1,\underline{r}_2) &= [q(\underline{r}_1,\underline{r}_2) - q_e(\underline{r}_1,\underline{r}_2)]/(1-|S_{ab}|^2)
\end{aligned}
$$

where

$$
\begin{aligned}
q(\underline{r}_1,\underline{r}_2) &= [|a(\underline{r}_1|^2\cdot|b(\underline{r}_2)|^2 + |a(\underline{r}_2)|^2\cdot|b(\underline{r}_1)|^2]/2 \ , \\
q_e(\underline{r}_1,\underline{r}_2) &= \mathrm{Re}\{a^*(\underline{r}_1)b(\underline{r}_1)b^*(\underline{r}_2)a(\underline{r}_2)\} \ .
\end{aligned}
\tag{2.6}
$$

Along with the two-electron density (2.6) one can introduce the one-electron density that characterizes the probability of finding one electron in a volume $dv(\underline{r})$ provided the position of the second electron is arbitrary. Making use of (2.6) we obtain

$$
\begin{aligned}
P_1(\underline{r}) &= \int P_2(\underline{r},\underline{r}')dv(\underline{r}') \ , \\
{}^1P_1(\underline{r}) &= [q'(\underline{r}) + q_e'(\underline{r})]/(1+|S_{ab}|^2) \ , \\
{}^3P_1(\underline{r}) &= [q'(\underline{r}) - q_e'(\underline{r})]/(1-|S_{ab}|^2)
\end{aligned}
$$

where

$$
\begin{aligned}
q'(\underline{r}) &= [|a(\underline{r})|^2 + |b(\underline{r})|^2]/2 \ , \\
q_e'(\underline{r}) &= \mathrm{Re}\{a^*(\underline{r})b(\underline{r})S_{ba}\} \ .
\end{aligned}
\tag{2.7}
$$

From (2.6) and (2.7) it is seen that for the orthogonal orbitals a and b, when $S_{ab} = 0$, the one-electron densities in the singlet and the triplet states coincide, ${}^1P_1(\underline{r}) = {}^3P_1(\underline{r})$, but the two-electron densities do not coincide and differ from each other by the value

$$
{}^1P_2(\underline{r}_1,\underline{r}_2) - {}^3P_2(\underline{r}_1,\underline{r}_2) = 2q_e(\underline{r}_1,\underline{r}_2) \ .
$$

This means that in the case of orthogonal a and b orbitals the singlet-triplet splitting results solely from the correlation of electron motion in these states. Indeed from (2.5) at $S_{ab} = 0$ we obtain

$$
\hbar J_{AB} = -2\langle a(1)b(2)|e^2/r_{12}|b(1)a(2)\rangle \ . \tag{2.8}
$$

We see that $J_{AB} \leqq 0$ since the integral in the right-hand side is the intrinsic energy of a charge distributed in space with a density $a^*(\underline{r})b(\underline{r})$ and therefore is positive. Hence for $S_{ab} = 0$ the triplet state lies below the singlet one, i.e., the interaction between A and B is of ferromagnetic nature. Such a ferromagnetic type of exchange interaction occurs, for instance, between electrons occupying the orbitals of one and the same atom which leads to the well-known Hund rule, according to which for a given electron configuration the ground state of an atom is the one with the maximum multiplicity allowed by the Pauli principle.

In the case of nonorthogonal orbitals when $S_{ab} \neq 0$ both two- (2.6) and one-electron (2.7) densities are different in the singlet and the triplet states. Therefore, in addition to the effect of interelectron interaction an essential contribution to the exchange integral is made by the electrostatic interaction between electrons and nuclei [the first four terms in (2.5)]. As a result the sign of the exchange integral may become positive, $J_{AB} > 0$. Precisely this antiferromagnetic type of exchange interaction is typical of interatomic interaction.

For the nonorthogonal orbitals all the terms in (2.5) are quantities of the same order of magnitued and are, roughly speaking, proportional to the square of the overlap integral, i.e.,

$$J_{AB}(\underline{r}) \sim |S_{ab}(\underline{r})|^2 \quad . \tag{2.9}$$

Since the overlap integral decreases exponentially as the distance between the atoms increases, so from (2.9) it is seen that the exchange integral is also an exponentially decreasing function of r.

Interaction Between Many-Electron Systems. Delocalization and Spin Polarization

Practically of the greatest interest is the interaction of paramagnetic particles containing many electrons rather than one. For such systems in the unrestricted Hartree-Fock approximation the exchange integral may be expressed as the sum [2.12]

$$J_{AB} = \sum_{a,b} J(a,b)/n_A \cdot n_B \tag{2.10}$$

in which each term describes the contribution of the pair of orbitals a and b of the interacting particles A and B while n_A and n_B are the number of unpaired electrons of particles A and B, respectively. In [2.13] such an orbital approximation was used to calculate the exchange integral for interaction between various atoms. At distances approximately equal to the sum of Van der Waals radii of atoms the decrease in J_{AB} with the growth of r proved to proceed even faster than follows from (2.9). According to [2.13] the correlation $J_{AB} = J_0 S_{ab}^n$ with n somewhat larger than 2 is well fulfilled in the region of Van der Waals distances. Figure 2.2 presents the dependences of the pair-orbital exchange integrals J(2pσ, 2pσ) and J(2pπ, 2pπ) on the distance between the atoms calculated in [2.13]. It can be seen that J decreases exponentially as r increases, the main contribution to J being made by the σ orbitals of the atoms. The values of J_0 and n [2.13] are given in Table 2.1.

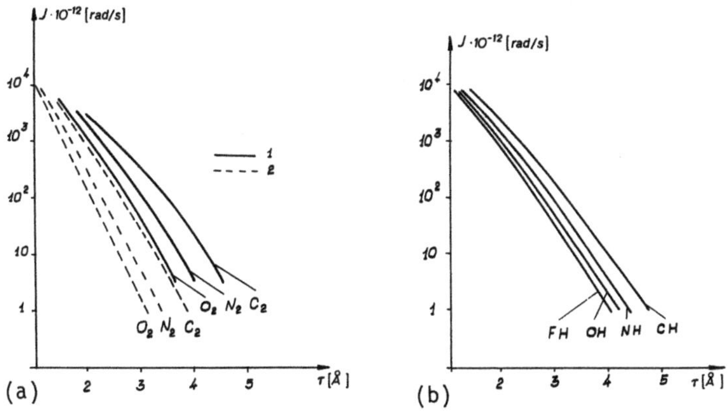

Fig. 2.2a,b. Exchange integrals in pair orbital approximation (a) 1 - J(2pσ, 2pσ), 2 - J(2pπ, 2pπ); (b) J(2pσ, 1s). The figure presents the dependencies of the exchange integrals on the interatomic distances for various types of the overlapping orbitals

Table 2.1. Parameters of approximation formula $J_{AB} = J_0 \cdot S_{ab}^n$ [2.13]

Atoms	H-H	Li-Li	B-B	C-C	N-N	O-O	F-F
$J_0 \cdot 10^{-16}$ [rad/s]	3.53	0.94	2.40	5.78	9.46	10.92	12.96
n	2.22	2.33	2.30	2.36	2.42	2.49	2.54

In polyatomic paramagnetic particles – free radicals, metal complexes, excited triplet molecules – unpaired electrons occupy molecular, rather than atomic, orbitals, i.e., are delocalized over a molecule. The delocalization of unpaired electrons is of great importance for the exchange interaction of paramagnetic particles since in most cases exchange interaction is mainly determined by the orbital overlap of those atoms of A and B that are in close contact. Therefore the scope of exchange interaction energies will greatly depend on the extent of delocalization of unpaired electrons onto the orbitals of the atoms in contact.

We illustrate the role of electron delocalization in exchange interaction by an example of interaction between an atom A and a paramagnetic complex B. Let a be the orbital of an unpaired electron of A and the unpaired electron of B be distributed between an orbital m of the central atom of the complex and an orbital b of the ligand, i.e., let it occupy the orbital $\psi = c_m \cdot m + c_B \cdot b$. In real systems the unpaired electron is most commonly concentrated on the central atom of the complex so that $|c_m|^2 \gg |c_b|^2$. Hence, it would seem, in exchange interaction the main part should be played by the direct overlap of the orbital of the central atom with that of the interacting partner. However the overlap of a with m is essentially less than that of a with the orbital of the ligand b. So in exchange interaction the leading role belongs to the delocalization of the unpaired electron onto the orbital of the ligand and to the overlap of a with b (Fig.2.3).

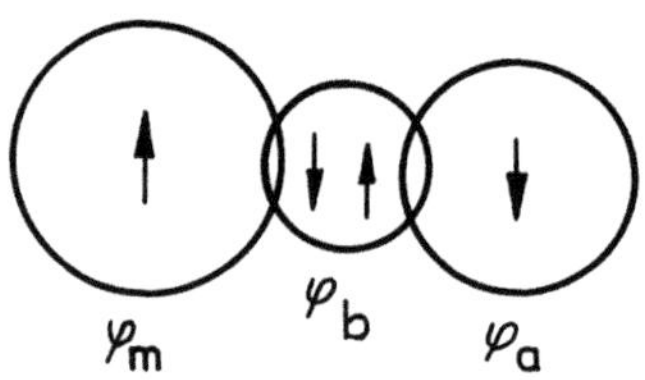

Fig. 2.3. Exchange interaction via a diamagnetic molecule. There is no direct overlap between ψ_a and ψ_m orbitals. However both orbitals overlap with the orbital ψ_b of the bridging diamagnetic molecule

Along with the effect of unpaired electron delocalization in the exchange interaction of many-electron paramagnetic particles, the effect of spin polarization may play a great role. The interaction of the unpaired electrons with the electrons of the filled shells results in the splitting of the molecular orbitals of the filled shells since the unpaired electron excites differently the filled orbitals occupied by electrons with equal spatial but different (α and β) spin states. In consequence of the spin polarization of the filled shells, an additional unpaired spin density appears on the atomic orbitals of the paramagnetic particles A and B so that the total spin density on the atoms of the A and B particles will contain a contribution from both the delocalization of the unpaired electrons and the polarization of the filled orbitals. In this situation the exchange interaction of paramagnetic particles now depends on the overlap of the orbitals of both the unpaired electrons themselves and of the filled shells of A and B perturbed by the unpaired electrons.

The interaction resulting from the overlap of orbitals of paramagnetic particles (a and m in the above example) is often called in the literature the direct exchange. The interaction induced by the delocalization of the unpaired electron onto the orbitals of the bridging diamagnetic molecule or by the polarization of orbitals of a diamagnetic particle is called the indirect exchange or superexchange [2.2, 14]. For paramagnetic particles screened by diamagnetic particles it is generally the indirect exchange which is the basic mechanism of exchange interaction [2.14]. Of course, there may also exist intermediate situations when the exchange interaction is appreciably contributed to by both the direct and the indirect mechanisms. For example, the exchange interaction of ions V^{2+} separated by diamagnetic ion F^- is shown to be essentially contributed to by the direct overlap of the orbitals of V^{2+} as well [2.16].

Anisotropy of Exchange Interaction

The overlap of the orbitals of unpaired electrons depends on the mutual orientation of two paramagnetic particles. Therefore the exchange interaction, too, depends on the angles characterizing the mutual orientation of the particles. For example, in [2.18] for the system $\dot{C}H_3$... H (Fig.2.4) the angular dependence of J was calculated (see Fig. 2.5). Depending on the mutual orien-

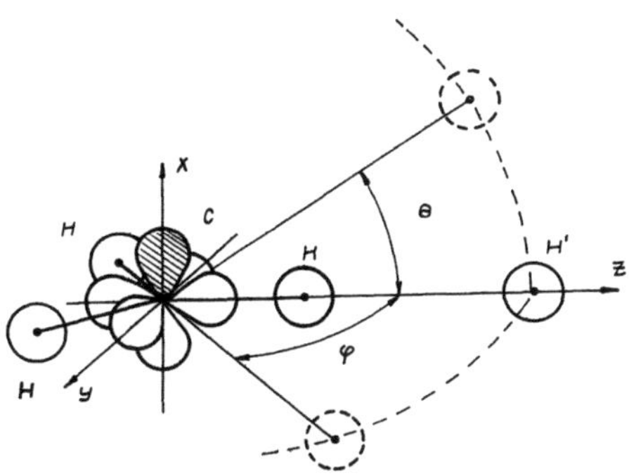

Fig. 2.4. Position of methyl radical relative to the hydrogen atom in the system $\dot{C}H_3$ ... H

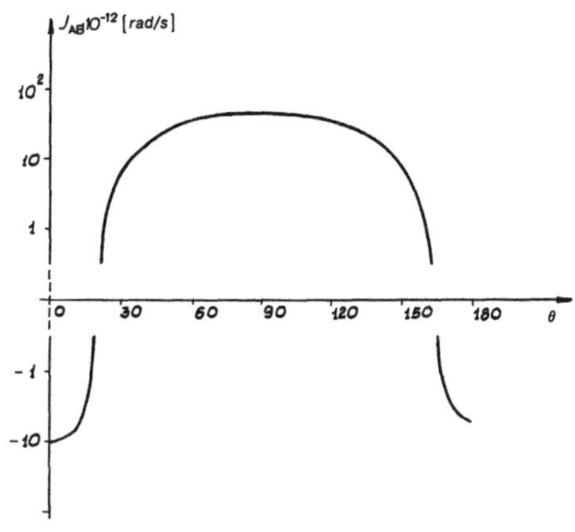

Fig. 2.5. Dependence of the exchange integral on the angle θ for the pair $\dot{C}H_3$ ... H at $r_{CH'}$ = 3.49 Å. Angle and distance designation are as in Fig. 2.4

tation of partners J even changes its sign. If H is located on the symmetry axis of $\dot{C}H_3$ the exchange integral is due to the direct overlap of the 1s orbital of the H atom with the 2p orbital of the unpaired electron of $\dot{C}H_3$. If H is on the horizontal plane of symmetry of $\dot{C}H_3$ the main contribution to the exchange interaction is due to the negative spin density on the hydrogen atoms of the methyl radical. Note that the negative spin density arises through the mechanism of spin polarization. The latter example serves as a good illustration of the correlation between the exchange integral and the spin density of atoms that are in close contact.

Semiempirical Estimates of Exchange Integral

Consistent theoretical calculations of the exchange integral have so far been made for only a few polyatomic paramagnetic species [2.15-18]. Therefore rather wide use is made of the semiempirical estimation of the exchange integral with the help of a correlation suggested by OWEN [2.19]. According to [2.19] the main contribution to the exchange interaction of two paramagnetic particles A and B is made by the overlap of orbitals of atoms a and b which happen to be in close contact. This assumption is supported by the fact that it is the indirect exchange which is often the basic mechanism of exchange interaction in magnetic dielectrics [2.4, 14] as well as by the qualitative considerations stated above. In the framework of this approximation the exchange interaction between A and B differs from zero only when on the atoms a and b that are in contact there is a nonzero spin density, i.e., there are uncompensated spins. For these reasons OWEN expressed the exchange integral in one of the following forms [2.19]:

$$J_{AB} = J_{ab} \cdot \rho_a \rho_b / n_A \cdot n_B \ ,$$

$$J_{AB} = \alpha J_{ab}(r_0) \rho_a \rho_b / n_A \cdot n_B \ . \qquad (2.11)$$

These expressions contain the following notation: J_{ab} is the exchange integral between atoms a and b at a Van der Waals distance between them: ρ_a and ρ_b are the spin densities on the atoms a and b; n_A and n_B are the numbers of unpaired electrons of A and B; $J_{ab}(r_0)$ is the exchange integral in a diatomic molecule ab with a chemical bond spacing r_0 between the nuclei; a dimensionless parameter α takes account of the decrease in the exchange integral due to the fact that when A and B come in contact the distance between the atoms a and b equals the sum of their Van der Waals radii rather than r_0. The value of $J_{ab}(r_0)$ is roughly equal to the doubled energy of dissociation of the ab molecule so that $J_{ab}(r_0) \sim 10^{16}$ rad/s. Using the Morse potential for interatomic interaction, OWEN suggested the following formula to find α:

$$\alpha = 1 - \{1 - \exp[-\varkappa(r - r_0)]\}^2$$

or

$$\alpha \approx 2 \exp[-\varkappa(r - r_0)] \quad \text{when } r \gg r_0 \quad . \qquad (2.12)$$

The parameter $\varkappa$ depends on the kind of atoms and amounts to about 10^8 cm^{-1} [2.20]. Substituting the sum of the Van der Waals radii of atoms a and b for r in (2.12) one can find the values of the parameter α for any pair of atoms. The results of these calculations [2.21] for some pairs of atoms are presented in Table 2.2 from which it can be seen that for different pairs of atoms the values of α vary no more than severalfold. The value of α can also be calculated from the data reported in [2.13] (see Fig. 2.2). For example, in a molecule of O_2 the length of the chemical bond is equal to 1.2 Å. From Fig. 2.2 it is seen that as the distance between atoms increases to become a Van der Waals one the exchange integral decreases by about two orders. The values of α found from the data of [2.13] are listed in Table 2.2. Calculated by different methods these values are nevertheless in good agreement. Typically at the Van der Waals distances between atoms we obtain for the exchange integral the values of $J_{ab} \approx 2 \cdot 10^{14}$ rad/s.

Table 2.2. Values of parameter α

Pair of atoms	Sum of Van der Waals radii [Å]	[2.21]	[2.13]
O ... O	2.8	0.02	0.013
O ... H	2.6	0.04	0.016
H ... Cl	3.0	0.07	-
C ... H	3.0	0.02	0.015
Cl... Cl	3.6	0.07	-

The exchange integral at the Van der Waals distances can also be estimated by quantum chemical calculations [2.13,15,17,18]. The results of such calculations for the exchange interaction of unpaired electrons on the 1s orbitals of hydrogen atoms (1s - 1s interaction) and for the 2pσ - 2pσ interaction in the case of fluorine, oxygen, and carbon atom pairs are given in Table 2.3 [2.13]. Within an order of magnitude they are equal to the estimates obtained with the help of the Morse potential.

Table 2.3. Values of the exchange integral for atom pairs at Van der Waals distances [2.13]

Pair of atoms	Distance between atoms [Å]	J_{ab} [rad/s]
H ... H	2.40	$3.5 \cdot 10^{14}$
O ... O	2.80	$1.1 \cdot 10^{14}$
C ... C	3.70	$0.6 \cdot 10^{14}$
F ... F	2.80	$0.4 \cdot 10^{14}$

The spin density on the atoms a and b can be measured directly by radiospectroscopic methods [2.22]. So correlations (2.11) are a really useful working tool for qualitative reasoning and quantitative estimation of the exchange integral, particularly for a comparative analysis of the exchange interaction within a series of similar paramagnetic particles.

Note that according to (2.11) [see also (2.10)] the parameter J_{AB} of the spin-Hamiltonian (2.1) is less than the exchange integral J_{ab} for a pair of overlapping orbitals by a factor of $n_A \cdot n_B$. The appearance of the factor $1/(n_A n_B)$ can be interpreted as follows. In the model in question the main role is played by the overlap of orbitals of only those atoms A and B that are in close contact; therefore the strength of the interaction of A and B is determined by the interatomic exchange integral J_{ab}. The strength of the exchange interaction betweeen the two atoms and, consequently, the strength of the interaction between A and B seems to be weakly dependent on the total number of unpaired electrons of A and B. At the same time the strength of the interaction can be characterized, according to (2.1), e.g., by the total splitting of the terms $E_{1+2S_B} - E_1 = \hbar J_{AB} S_B(1 + 2S_A)$ at $S_B \leq S_A$. The spin factor on the right-hand side of the formula increases as the number of unpaired electrons of A and B increases since $n_A = 2S_A$, $n_B = 2S_B$. It is natural, therefore, that the effective parameter J_{AB} should fall as $n_A \cdot n_B$ grows, just as expressed in (2.11).

Let us consider to what extent Owen correlation agrees with the quantum chemical picture of the mechanism of spin density delocalization.

When delocalization is responsible for the spin density expansion into the peripheral sites of contact of partners, the Owen correlation is equivalent to the above formula (2.9). Accordingly the exchange integral is proportional to the square of the orbital overlap integral of the unpaired electrons of A and B. This is illustrated by the diagram of indirect exchange (see Fig.2.3). The orbital overlap integral of the unpaired electrons of A and B equals $S_{AB} = c_m S_{am} + c_b S_{ab}$. Neglecting the direct overlap of the orbital of A with that of the central ion we come to the result $S_{AB} \approx c_b S_{ab}$. Keeping terms up to second order in S_{AB} [see (2.9)] we obtain $J_{AB} \sim |c_b S_{ab}|^2$. Thus we arrive at the conclusion that the exchange integral is proportional to $|c_b|^2$, i.e., to the density of the unpaired electron at the point of contact.

The situation becomes much more complicated if a notable spin polarization occurs [2.12, 23]. In [2.23] an analysis was made of the connection of the exchange integral with the distribution of the spin density of interacting paritcles on the basis of a three-centre four-electron model suggested in the theory of superexchange [2.24]. It is shown [2.23] that the exchange integral of the four-electron system (see Fig.2.3) can be represented as the sum of contributions from the delocalization of an unpaired electron at the point of contact [ρ_b(deloc.)] and from the spin polarization in a contacting atom [ρ_b(polar.)]

$$J_{AB} = J'_{ab}\rho_a\rho_b(\text{deloc.}) + J''_{ab}\rho_a\rho_b(\text{polar.}) \quad . \tag{2.13}$$

As is seen, the dependence of the exchange integral on the spin densities in the general case is more complex than follows from the correlations (2.11). However in two limiting and practically important cases when one of the terms on the right-hand side of (2.13) is much greater than the other, i.e., when one of the mechanisms (delocalization or spin polarization) is predominant, the exchange integral can be written in the form

$$J_{AB} \sim \rho_a\rho_b \ , \tag{2.14}$$

where ρ_a and ρ_b are the total spin densities on the atoms a and b that can be measured in experiment. Under certain conditions, however, correlation (2.14) may prove valid even when both mechanisms work effectively. The latter will take place when the spin densities ρ(deloc.) and ρ(polar.) are concentrated on one and the same outer atomic orbital of the atom b. In this case in (2.14) $\rho_b = \rho_b$(deloc.) + ρ_b(polar.). Consequently, in many real situations the application of (2.11) can be justified.

An experimental proof of the possibility of using (2.11) for estimating the exchange integral was reported in [2.19]. There this equation was used to analyze the data for crystal salts of iridium in which a pair of iridium ions is bound by a bridge of chlorine or bromine anions

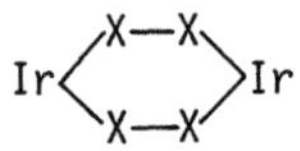
, X = Cl, Br .

For these compounds, the values of the exchange integrals between iridium atoms were obtained by measuring the magnetic susceptibility. In addition, by means of ESR spectroscopy the density of the unpaired electron on each of the bridging anions X was found to be about 5%. Table 2.4 compares the values of the exchange integral for iridium salts calculated from (2.11) with those measured experimentally. The comparison confirms that (2.11) does allow one to obtain a satisfactory estimate of the exchange integral.

Table 2.4. Values of the exchange integral for iridium salts [2.19]

Salt	Ir-Ir [Å]	Cl-Cl [Å]	J [rad/s] observed	J [rad/s] calculated
Cs_2IrCl_6	7.24	3.74	$0.14 \cdot 10^{12}$	$0.59 \cdot 10^{12}$
Rb_2IrCl_6	7.00	3.50	$0.66 \cdot 10^{12}$	$0.91 \cdot 10^{12}$
$(NH_4)_2IrCl_6$	6.98	3.48	$0.94 \cdot 10^{12}$	$0.96 \cdot 10^{12}$
K_2IrCl_6	6.90	3.40	$1.5 \cdot 10^{12}$	$1.12 \cdot 10^{12}$
		Br-Br [Å]		
K_2IrBr_6	7.25	3.57	$1.8 \cdot 10^{12}$	$1.3 \cdot 10^{12}$

Note that more rigorous theoretical treatments also confirm the applicability of the semiempirical relation (2.11). For example, calculations of the exchange integral for the interaction of the fragment-$\dot{C}H$- with the allyl radical [2.12] and of the atom H with the methyl radical [2.18] based on (2.11) yield values of J_{AB} in good agreement with results of quantum chemical calculations [2.5]. The existence of a correlation between the exchange integral and the spin density has also been confirmed by the calculation of the interaction between the ions of V^{2+} and those of Ni^{2+} in MgO [2.25].

2.2 Qualitative Description of Spin Exchange Process

Spin Exchange

In most cases the exchange interaction of two paramagnetic particles that have approached each other to a distance equal to the sum of Van der Waals radii is small compared to their heat energy: $J_{AB} \lesssim kT\hbar^{-1} \approx 10^{13}$ rad/s. This interaction, therefore, does not practically affect the thermal motion of molecules in a solution. But it can essentially change the state of the spins of paramagnetic particles.

The change of the spin state of colliding particles caused by exchange interaction (2.1) usually, though not always, is nothing else but an exchange of spin states between the partners, i.e., a change in the state of electron spin of one of the particles is fully compensated by an opposite change in the electron spin state of its partner. This follows from the fact that (2.1) does not change the total spin moment of both partners. Indeed, the spin-Hamiltonian (2.1) commutes with the operators of the total spin projections $\hat{S}_u = \hat{S}_{uA} + \hat{S}_{uB}$, $u = x,y,z$, i.e., $[\hat{S}_u,\hat{V}] = 0$. Hence, taking account of the equation of motion [2.6]

$$\partial\hat{S}_u/\partial t = -i\hbar^{-1}[\hat{S}_u,\hat{V}] \tag{2.15}$$

we find that for exchange interaction (2.1) the total spin is an integral of motion, $\hat{S}_u$ = const, while the changes in the projections of the partners' spin moments have opposite signs,

$$(\partial/\partial t)\hat{S}_{Au} = -(\partial/\partial t)\hat{S}_{Bu} \quad . \tag{2.16}$$

In real systems within the time t_c of contact of paramagnetic particles the motion of their spins is, as a rule, controlled only by exchange interaction (2.1). And then the collision of particles leads to an exchange of spin states between the partners, i.e., to a spin exchange in the true sense of the word.

It should be noted that from here on the term "spin exchange" will be used in a wider sense. It will be applied to any spin state changes of colliding particles caused by their exchange interaction including those resulting not only in an equivalent spin state exchange between the partners but also in some other changes in those states. Such a situation may arise when the collision time is long enough both for the exchange interaction and other spin-dependent interactions of unpaired electrons (hyperfine interaction with magnetic nuclei, spin-orbital interaction, zero field splitting energy, etc.) to reveal themselves. These processes may interfere in a fairly complicated fashion with the exchange interaction. A specific example of this interference effect in spin exchange will be considered at the end of this section.

Spin Exchange in a Two-Electron System

An example of the elementary process of spin exchange is a collision of two atoms A and B either of them having one electron. This example will permit us to illustrate most clearly the essence of the elementary act of spin exchange. Assume that before they collided the electrons had been in a state with the wave function

$$\psi_i = (1/\sqrt{2}) \cdot \mathrm{Det}|a\alpha, b\beta| \; , \tag{2.17}$$

i.e., the positive spin density had been concentrated on the atom A (α is the electron spin orientation on the orbital a) and the negative one on the atom B (β is the electron spin orientation on the orbital b). After the atoms collide the electrons may either remain in the same state ψ_i (elastic collision) or change their spin states, i.e., the system will pass into an alternative state with the negative spin denisty on the A atom and the positive spin density on the B atom (inelastic collision). The wave function of this state is

$$\psi_f = (1/\sqrt{2}) \cdot \mathrm{Det}|a\beta, b\alpha| \; . \tag{2.18}$$

How does the turning on of the exchange interaction at the collision of A and B lead to a spin exchange? When the distance between the atoms is large and the overlap of their electron shells is negligible the wave functions ψ_i (2.17) and ψ_f (2.18) describe the eigenstates of the system betweeen which no transition occurs. But as soon as the atoms A and B contact the situation changes: the electron orbitals a and b overlap and an exchange of spin states between A and B becomes feasible. The change of the wave function during the contact of the atoms can be determined by solving the Schroedinger equation [2.6]

$$i\hbar\partial\psi/\partial t = \hat{H}_e\psi \tag{2.19}$$

where $\hat{H}_e$ is the electron Hamiltonian of system (2.4). To solve this problem it is convenient to introduce a new basis of the singlet and triplet states of the system. The wave functions of the singlet and the triplet are equal to [2.11]

$$\psi_S = c_S[a(1)b(2) + a(2)b(1)] \times [\alpha(1)\beta(2) - \beta(1)\alpha(2)] ,$$

$$\begin{pmatrix} \psi_{T_+} \\ \psi_{T_0} \\ \psi_{T_-} \end{pmatrix} = c_T[a(1)b(2) - a(2)b(1)] \times \begin{pmatrix} \alpha(1)\alpha(2) \\ [\alpha(1)\beta(2) + \beta(1)\alpha(2)]/\sqrt{2} \\ \beta(1)\beta(2) \end{pmatrix} . \tag{2.20}$$

Here T_+, T_0, T_- stand for the sublevels of the triplet with the projections of the total spin which are equal to +1,0,-1, respectively; c_S, c_T are normalizing factors.

The wave function of the system before a collision when $S_{ab} = 0$ can be written in the form of a linear combination of ψ_S and ψ_T

$$\psi_i = (\psi_S + \psi_{T_0})/\sqrt{2} . \tag{2.21}$$

Exchange interaction does not change the multiplicity of the system; therefore the statistical weight of S and T_0 in the wave function (2.21) is not changed during the collision. But due to a difference in the energy of S and T states the phases of the wave function ψ_S and ψ_{T0} do not change in the same manner, namely according to (2.19) we have

$$\begin{aligned} \psi(t) &= (1/\sqrt{2}) [\psi_S \exp(-iE_St/\hbar) + \psi_{T_0} \exp(-iE_Tt/\hbar)] \\ &= (1/\sqrt{2}) [\psi_S \exp(iJt) + \psi_{T_0}] \cdot \exp(-iE_Tt/\hbar) . \end{aligned} \tag{2.22}$$

Hence we find that after the collision the wave function of the system $\psi(t_c)$ may be presented as a combination of ψ_i and ψ_f

$$\psi(t_c) = \psi_i \cdot \cos(Jt_c/2) - i\psi_f \cdot \sin(Jt_c/2) , \tag{2.23}$$

where t_c is the collision duration. In deriving (2.23) from (2.22) the factor $\exp[-i(E_S + E_T)t_c/2\hbar]$ was neglected.

Thus, from (2.23) it can be seen that during the collision, the atoms undergo a spin exchange with the probability

$$p(t_c) = \sin^2(Jt_c/2) \quad . \tag{2.24}$$

If $Jt_c = \pi$, then such a collision completely changes the sign of the spin density of the partners. This process which consists in a simultaneous changing of the spin density sign on A and B is referred to as a flip-flop process.

As long as we are interested only in the spin state of electrons on atoms A and B, we may confine ourselves to indicating only the spin states of electrons rather than describing in detail their entire wave function. The transitions between the states $\psi_i = (1/\sqrt{2})\ \mathrm{Det}|a\alpha,b\beta|$ and $\psi_f = (1/\sqrt{2})\ \mathrm{Det}|a\beta,b\alpha|$ may be regarded as transitions between the two spin states $\alpha_A\beta_B$ and $\beta_A\alpha_B$ of atoms A and B without indicating the orbitals a and b occupied by the electrons. The spin state changes of A and B when these collide can then be readily considered with the help of the spin-Hamiltonian of exchange interaction (2.1). Reducing the study of the processes brought about by exchange interaction to a pure spin problem does not mean that we are no longer interested in the structure of electron orbitals. The direct influence of a particular kind of a and b on the process of spin exchange finds its accumulated expression through the value of the exchange integral - the parameter of the spin-Hamiltonian of exchange interaction.

Spin Density Matrix

The spin motion can be most conveniently described using the formalism of the density matrix [2.6]. The spin state of a certain paramagnetic particle with one unpaired electron can be presented in the form of a linear combination of α and β states corresponding to the two eigenvalues of spin projections

$$\psi = c_1\alpha + c_2\beta \ , \quad |c_1|^2 + |c_2|^2 = 1 \quad . \tag{2.25}$$

The wave function (2.25) describes completely the state of the spin in question. With its help one can calculate the value of any spin characteristic Q

$$Q = \langle\psi|\hat{Q}|\psi\rangle = |c_1|^2 Q_{\alpha\alpha} + c_1 c_2^* Q_{\alpha\beta} + c_1^* c_2 Q_{\beta\alpha} + |c_2|^2 Q_{\beta\beta} \quad . \tag{2.26}$$

The matrix ρ made up of the coefficients c_1 and c_2 ,

$$\rho = \begin{pmatrix} |c_1|^2 & c_1^* c_2 \\ c_1 c_2^* & |c_2|^2 \end{pmatrix} \equiv \begin{pmatrix} \rho_{\alpha\alpha} & \rho_{\alpha\beta} \\ \rho_{\beta\alpha} & \rho_{\beta\beta} \end{pmatrix} , \quad \rho_{\alpha\beta} = \rho_{\beta\alpha}^* , \tag{2.27}$$

is called the spin density matrix. The diagonal elements of ρ, $\rho_{\alpha\alpha} = |c_1|^2$, and $\rho_{\beta\beta} = |c_2|^2$ are real; they characterize the population of the states with α and β orientations of the spin, respectively. The nondiagonal elements are, in the general case, complex quantities; they assign a spin projection on a plane perpendicular to the axis of quantization.

By using ρ (2.26) can be written as

$$Q = \mathrm{Tr}\{\rho\hat{Q}\} \equiv \sum_{n,m} \rho_{nm} Q_{mn} \quad . \tag{2.28}$$

Let us choose the axis of quantization to be the Z axis. Then α and β are the eigenfunctions of the operator $\hat{S}_z$ that correspond to eigenvalues +1/2 and -1/2. The mean values of the spin projections are expressed via the density matrix elements as follows:

$$\begin{aligned} S_x &= \mathrm{Tr}\{\rho\hat{S}_x\} = (\rho_{\alpha\beta} + \rho_{\beta\alpha})/2 = \mathrm{Re}\{\rho_{\alpha\beta}\} , \\ S_y &= \mathrm{Tr}\{\rho\hat{S}_y\} = \mathrm{Im}\{\rho_{\beta\alpha}\} , \\ S_z &= \mathrm{Tr}\{\rho\hat{S}_z\} = (\rho_{\alpha\alpha} - \rho_{\beta\beta})/2 , \end{aligned} \tag{2.29}$$

where Re and Im stand for the real and the imaginary parts, respectively. Let us write the complex number $\rho_{\alpha\beta}$ in the form $\rho_{\alpha\beta} = |\rho_{\alpha\beta}| \cdot \exp(-i\theta)$; then

$$\begin{aligned} S_x &= |\rho_{\alpha\beta}| \cdot \cos\theta , \\ S_y &= |\rho_{\alpha\beta}| \cdot \sin\theta \quad . \end{aligned} \tag{2.30}$$

From the latter two formulae it is seen that the difference of the diagonal elements of ρ determines the longitudinal polarization of the spin (along the Z axis) and the nondiagonal element characterizes the transversal projections of the spin. The phase of the nondiagonal element θ equals the angle which the X axis forms with the spin projection on plane XY.

The experiment involves the measurement of the characteristics of an ensemble of spins rather than of an individual one. For a comparison with the experiment, therefore, relations (2.26,28) must be averaged over the ensemble. It is at this stage of statistical averaging that the advantages of the density matrix over the wave function are revealed. The wave function is not a directly measured quantity so it cannot be averaged over an ensemble of spins. As for the elements of the density matrix, they are measurable directly and can be averaged over all the spins. The density matrix averaged with respect to an ensemble of N spins is equal to

$$\rho = \begin{pmatrix} \sum_k \rho_{\alpha\alpha}^{(k)}/N & \sum_k \rho_{\alpha\beta}^{(k)}/N \\ \sum_k \rho_{\beta\alpha}^{(k)}/N & \sum_k \rho_{\beta\beta}^{(k)}/N \end{pmatrix} ,$$

where the summation is performed over all the spins. In the following discussion ρ will stand everywhere for a statistically averaged density matrix.

Spin Motion Due to Exchange Interaction

The time evolution of ρ is given by [2.6]

$$\partial\rho/\partial t = -i\hbar[\hat{H},\rho] , \tag{2.31}$$

where $\hat{H}_e$ is the spin-Hamiltonian of the system. The spin-Hamiltonian of a pair of paramagnetic particles, along with the exchange interaction, includes some others: dipole-dipole, hyperfine, etc. As a rule, spin exchange processes are studied by the methods of magnetic spectroscopy, i.e., under the conditions when the spins are placed into an external steady magnetic field. Therefore $\hat{H}_e$ also includes the Zeeman energy of the spins in the external mangetic field. Let us present $\hat{H}$ as the sum of two terms,

$$\hat{H} = \hat{H}_0 + \hbar J\hat{S}_A\hat{S}_B , \tag{2.32}$$

where $\hat{H}_0$ contains all the spin interactions except the exchange one.

The character of the spin exchange process depends on a particular kind of $\hat{H}_0$. If $\hat{H}_0$ commutes with the exchange Hamiltonian, then the spins of A and B perform two independent motions: one under the action of $\hat{H}_0$ and the other under that of exchange interaction. An example of such a situation may be the spin exchange between two particles with identical Larmor frequencies. For such a system, by choosing the direction of the external magnetic field to be the Z axis and denoting the Larmor frequency by ω_0 we obtain

$$\hat{H}_0 = \hbar\omega_0(\hat{S}_{Az} + \hat{S}_{Bz}) . \tag{2.33}$$

One can make certain that for (2.33)

$$[\hat{H}_0, \hbar J\hat{S}_A\hat{S}_B] = 0 . \tag{2.34}$$

In the case under discussion the exchange interaction produces an equivalent exchange of the partners' spin states [see (2.16)], and a flip-flop process occurs.

Let us consider the motion of spins in more detail.

Substituting (2.33) into (2.32) and solving the equation of motion (2.31) we obtain

$$\rho(t) = \hat{L}(t)\rho(0)\hat{L}^{-1}(t) , \quad (2.35)$$

where $\rho(0)$ is the initial density matrix and the evolution operator is equal to

$$\hat{L}(t) = \exp(-i\hat{H}t/\hbar) = \exp(-i\hat{H}_0 t/\hbar) \cdot \exp(-iJ\hat{S}_A\hat{S}_B t) . \quad (2.36)$$

In the basis of the eigenfunctions of operators $\hat{S}_{Az}$ and $\hat{S}_{Bz}$ equal to

$$|1\rangle = \alpha_A\alpha_B, \ |2\rangle = \alpha_A\beta_B , \ |3\rangle = \beta_A\alpha_B, \ |4\rangle = \beta_A\beta_B \quad (2.37)$$

the matrix elements of $\hat{L}$ are equal to

$$L_{11} = \exp(-i\omega_0 t - iJt/4) ,$$
$$L_{33} = \cos(Jt/2) \cdot \exp(iJt/4) ,$$
$$L_{44} = \exp(i\omega_0 t - iJt/4) ,$$
$$L_{23} = L_{32} = -i \sin(Jt/2) \cdot \exp(iJt/4) . \quad (2.38)$$

The other matrix elements equal zero.

Assume that at the initial moment the spin of A is in an α state, and that of B in a β state, i.e., $\rho(0)$ has the only element $\rho_{22}(0) = 1$ that is not equal to zero.

From equations (2.36,38) we find that

$$\rho_{22}(t) = |L_{22}|^2 = \cos^2(Jt/2) ,$$
$$\rho_{33}(t) = |L_{23}|^2 = \sin^2(Jt/2) ,$$
$$\rho_{23}(t) = \rho_{32}^*(t) = L_{22}L_{23}^* = (i/2) \sin(Jt) . \quad (2.39)$$

With the help of (2.39) and (2.29) one can find out how the projections of the partners' spins vary with time. The transverse spin projections remain equal to zero. This is due to the fact that for a chosen initial state the phases of spins θ_A and θ_B [see (2.30)] are distributed with equal probability in the interval $(0,2\pi)$. As for the longitudinal spin components, the exchange interaction exercises a transfer of polarization between the partners

$$S_{Az} = 1/2 - \sin^2(Jt/2) ,$$

$$S_{Bz} = -1/2 + \sin^2(Jt/2) \quad . \tag{2.40}$$

The exchange of Z projections between the spins leads to a flip-flop of the spins. The result obtained is equivalent to (2.24). It is of interest to note that for the spins in a magnetic field the exchange of the longitudinal polarization induces an exchange of the Zeeman energy between the spins. The flip-flop of the spins leads to an equivalent exchange of their transverse components as well.

The exchange of the partners' spin states ceases to be equivalent, however, if $\hat{H}_0$ does not commute with the exchange interaction. We encounter such a situation when considering a pair of paramagnetic particles with different Larmor frequencies. The spin-Hamiltonian takes the form

$$\hat{H}_0 = \hbar(\omega_A\hat{S}_{Az} + \omega_B\hat{S}_{Bz}) \quad . \tag{2.41}$$

In this case probability of a spin flip-flop resulting in an exchange of longitudinal polarization is equal not to $\sin^2(Jt/2)$ [see (2.39,40)], but to

$$p = \{J^2/[J^2 + (\omega_A-\omega_B)^2]\} \sin^2 \{[J^2 + (\omega_A-\omega_B)^2]^{\frac{1}{2}}t/2\} \quad . \tag{2.42}$$

If the difference of frequencies is large as compared with J the probability of a spin flip-flop is small, i.e.,

$$p \cong J^2/(\omega_A-\omega_B)^2 < 1 \quad .$$

This is due to the fact that at small values of J the exchange interaction cannot compensate the Zeeman energy changes of the spins performing the flip-flop process. But even the weak interactions when $|J| < |\omega_A-\omega_B|$ can essentially change the phase of colliding spins. Indeed, the exchange field of a collision partner is added up with the external field so within the collision time the spins acquire an additional phase. In this case the phase change of spins during a collision does not imply an equivalent exchange of the transverse components of the partners' spins as in the alternate case of $|J| < |\omega_A-\omega_B|$ but takes the form of spin dephasing. By way of illustration let us consider one example.

Assume the spin of A to be at the initial moment in a state α and that of B in a state $(\alpha + \beta)/\sqrt{2}$, i.e.,

$$\langle S_{Ax}(0)\rangle = \langle S_{Ay}(0)\rangle = 0 \ , \quad \langle S_{Az}(0)\rangle = 1/2 \ , \quad \langle S_{Bx}(0)\rangle = 1/2 \ ,$$
$$\langle S_{By}(0)\rangle = \langle S_{Bz}(0)\rangle = 0 \quad .$$

Solving the equation of motion (2.31) with the Hamiltonian (2.41) in the limiting case $|J| < |\omega_A - \omega_B|$ we obtain the following results. To an accuracy of negligibly small values of the order of $J^2/(\omega_A - \omega_B)^2$ the transverse components of the spin of A remain equal to zero $\langle S_{Ax}(t)\rangle \cong \langle S_{Ay}(t)\rangle \cong 0$ while those of the spin of B vary according to

$$\langle S_{Bx}(t)\rangle \cong (1/2) \cdot \cos(\omega_B t + Jt/2) \; ,$$
$$\langle S_{By}(t)\rangle \cong (1/2) \cdot \sin(\omega_B t + Jt/2) \; . \qquad (2.43)$$

As distinct from the situation of equivalent exchange [see (2.16)] the summarized projections of both spins $\langle S_{Ax}(t)\rangle + \langle S_{Bx}(t)\rangle$ and $\langle S_{Ay}(t)\rangle + \langle S_{By}(t)\rangle$ are not conserved. Hence it follows that in the case under consideration, as a result of the collision, the phase of B changes due to the interaction by a value $\Delta\theta_B = Jt/2$, while that of A remains uncertain. Therefore this collision simply displaces the precession phase of the spin of B.

In the ensemble of spin pairs there is a distribution with respect of the collision duration and the values of J. As a result, for the situation in question the exchange interaction produces a dephasing of spins - destroys the coherence of their precession. This is shown schematically in Fig. 2.6.

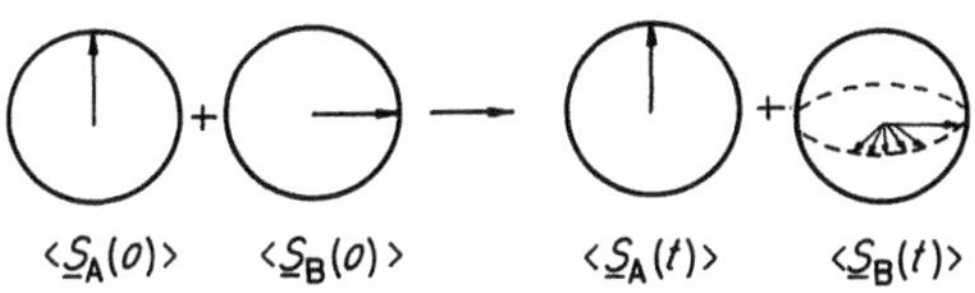

Fig. 2.6. Dephasing of spins of B. The phases of precession for various B spins are different due to the distributions over the values of exchange integrals and collision durations. As a result the coherency of precession is lost

Thus, spin exchange processes during a collision of two paramagnetic particles involve exchange of longitudinal spin projections (flip-flop), exchange of transverse spin projections (exchange of phases), and spin dephasing. Besides, the exchange of longitudinal and that of transverse spin projections are in fact one and the same process and occur at the same rate. This process of exchange of all spin projections which we call the equivalent exchange plays a dominant role, as a rule. As to the process of spin dephasing, it will be shown (Sect. 2.5) to be usually inessential.

2.3 Formal Kinetics of Spin Exchange in Magnetically Dilute Solutions

This book deals with spin exchange in magnetically dilute systems where the kinetics of the process is mainly controlled by binary collisions of paramagnetic species. Magnetically dilute systems have been the subject of numerous studies. Suffice it to say that most of the works on ESR have been done for such systems. Diluted solutions allow one to find out individual properties of paramagnetic particles and to study effects resulting from their binary collisions. In magnetically concentrated systems paramagnetic particles already form a unique interacting system and an important role in the spin dynamics is played by its collective properties. The latter are responsible for some of the most essential phenomena such as ferro-, ferri-, and antiferromagnetism, etc. There are quite a number of references summarizing the results of research into the magnetically concentrated systems [2.4].

The present section deals with the formal kinetics of spin exchange processes caused by binary collisions. Although this theory is purely phenomenological and does not allow the spin exchange rate constants to be expressed through the exchange integral and the kinetic parameters of thermal molecular motion, it proves sufficient to describe all the main spectroscopic manifestations of spin exchange. A more rigorous theory based on the analysis of microscopic equations for spin motion and virtually substantiating the validity of the suggested phenomenological consideration will be set forth in Sects. 2.5 and 2.6.

Binary Collision Approximation

Exchange interaction decays sharply, in accordance with the exponential law, with the growth of the distance r between paramagnetic particles. This short-range character of interaction makes it possible to introduce the concept of a spin exchange region with an effective radius r_i beyond which the spin interaction is negligible. The approximation of binary collisions is justified for such concentrations of spins C when the probabilities of three-particle, four-particle, etc., encounters are negligible. The probability of n collision partners getting simultaneously into the interaction region is equal to (ignoring the volume of the particles)

$$p = (1/n!)(v_iC)^n \cdot \exp(-v_iC) \ , \quad v_i = 4\pi r_i^3/3 \quad .$$

From here we see that the approximation of binary collisions is valid when conditions (2.44) is fulfilled,

$$v_iC < 1 \quad . \tag{2.44}$$

The value of the effective interaction radius r_i may be estimated as follows. The efficiency of spin exchange is determined by the action q performed within the collision time by the exchange interaction, $q = \hbar \int J(t)dt$. The estimates in the second order of the perturbation theory show that with a 1 percent accuracy the spin interaction beyond the r_i radius region can be neglected if $q(r > r_i)$ does not exceed $0.1\hbar$. Let us assume the exchange integral to decrease exponentially as r increases:

$$J(r) = J(R_{AB}) \cdot \exp[-\varkappa(r-R_{AB})] \ , \quad r \geq R_{AB} \ , \tag{2.45}$$

where $R_{AB} = R_A + R_B$ is the sum of the Van der Waals radii of the partners; $\varkappa$ characterizes the decrease of the exchange integral with increasing r. If we consider a central collision of A and B and assume the motion to be described by $r - R_{AB} = vt$ (v is the velocity of the relative motion of the partners), then for $q(r > r_i)$ we obtain

$$q(r > r_i) = \hbar J(R_{AB})/(\varkappa v) \quad . \tag{2.46}$$

Hence the radius of the interaction region can be estimated from the condition

$$J(r_i) \simeq 0.1 \ \varkappa v \ . \tag{2.47}$$

The estimates show that $r_i \cong R_{AB}$ and from (2.44) it follows that up to the concentrations $C \sim 10^{20}\ \mathrm{cm}^{-3}$ the spin exchange processes can be dealt with within the framework of binary collision approximation.

Rate of Bimolecular Spin Exchange Process

In diluted solutions the spin exchange processes can be described by means of bimolecular rates [2.27].

The spin exchange rate constant K_e is defined as the product of rate constant of diffusion encounters[3]

---

3 The value of K_D depends on the kinematics of the Brownian motion of particles. Formula (2.48) is valid when the displacement λ of particles in diffusion step is less than r_i. This situation appears to be typical of a liquid. If $\lambda > r_i$, then (2.48) overestimates the rate of diffusion encounters. This point is discussed in more detail in [2.26].

$$K_D = 4\pi D r_i \tag{2.48}$$

(where $D = D_A + D_B$ is the sum of the diffusion coefficients of the collision partners) and the probability p of a spin exchange in case the partners get into the sphere of interaction,

$$K_e = pK_D \quad . \tag{2.49}$$

One can introduce an effective radius of strong exchange r_e and assume that when the spins approach each other to a distance r_e, a spin exchange process necessarily occurs. Then the rate constant of spin exchange will equal

$$K_e = 4\pi D r_e \quad . \tag{2.50}$$

Exchange interaction and hence the efficiency of spin exchange in a collision depend on the mutual orientation of the partners' electron orbitals. Therefore a part of collisions may leave the state of the spins unchanged. The anisotropy of interaction is usually taken into account by introducing a steric factor f equal to the statistical weight of collisions which involve exchange interaction and in which spin exchange occurs. With allowance for the steric factor

$$K_e = fpK_D \quad . \tag{2.51}$$

The relaxation of mutual orientation of the partners averages the anisotropy of spin exchange, and as a result, the steric factor in (2.51) assumes the sense of an effective parameter.

Kinetics of Spin Exchange for Particles with Spins $S = 1/2$

One can investigate experimentally the spin exchange between both unlike and like paramagnetic particles. The former type includes spin exchange between free radicals and complexes of paramagnetic ions, between excited triplet molecules and those of oxygen, etc. When we speak about the spin exchange between like particles, say, free radicals or triplet molecules, what is meant is the spin exchange between particles belonging to different components of the hyperfine or fine structure of the ESR absorption spectra of these particles.

Let us consider the kinetics of spin flip-flop in the simplest case of a solution containing two kinds of particles with different g values and with spins S_A and S_B both equal to 1/2. Denote the density matrices and the concentrations of spins A(B) by $\rho^A(\rho^B)$ and $C_A(C_B)$, respectively. The concentration of spins A in the state α, $C_A\rho^A_{\alpha\alpha}$, decreases due to flip-flops with the spins B, i.e., due to the transitions $\alpha_A\beta_B \rightarrow \beta_A\alpha_B$, and increases due to collisions in which the process is inverted, $\beta_A\alpha_B \rightarrow \alpha_A\beta_B$. As a result, the total change in the concentration of particles A in the spin state α can be described by the conventional kinetic equation for bimolecular processes (see Fig. 2.6)

$$C_A\partial\rho^A_{\alpha\alpha}/\partial t = -K_eC_AC_B(\rho^A_{\alpha\alpha}\rho^B_{\beta\beta} - \rho^A_{\beta\beta}\rho^B_{\alpha\alpha}) \ , \tag{2.52}$$

where K_e is the rate constant of the bimolecular flip-flop process. Equations similar to (2.52) hold for the concentration of particles A in the state of the opposite spin orientation and for the concentration of particles B in the states α and β.

If the system under investigation is close to equilibrium, as is commonly the case in experiments on spin exchange in solutions, then (2.52) can be linearized. Let us represent ρ as the sum of an equilibrium density matrix ρ_0 and a matrix σ which takes account of the deviation from equilibrium

$$\rho = \rho_0 + \sigma \quad . \tag{2.53}$$

Substituting (2.53) into (2.52) and considering that as a rule, in real experiments on spin exchange, the high-temperature approximation

$$\rho_{0_{\alpha\alpha}} \cong \rho_{0_{\beta\beta}} \cong 1/2 \tag{2.54}$$

is valid, we obtain the linearized kinetic equation

$$C_A\partial\sigma^A_{\alpha\alpha}/\partial t = -(1/2)K_eC_AC_B[(\sigma^A_{\alpha\alpha}-\sigma^A_{\beta\beta}) - (\sigma^B_{\alpha\alpha}-\sigma^B_{\beta\beta})] \quad .$$

Similar equations can also be written for $C_A\sigma^A_{\beta\beta}$, $C_B\sigma^B_{\alpha\alpha}$, $C_B\sigma^B_{\beta\beta}$. These linearized equations assume an especially simple form, if we introduce the quantities

$$n_A = C_A(\sigma^A_{\alpha\alpha} - \sigma^A_{\beta\beta}) \ ,$$
$$n_B = C_B(\sigma^B_{\alpha\alpha} - \sigma^B_{\beta\beta}) \tag{2.55}$$

which give a nonequilibrium difference of the concentrations of spins with α and β orientation. The variations of n_A and n_B due to the bimolecular spin flip-flop are described by the equations

$$\partial n_A/\partial t = -K_e C_B n_A + K_e C_A n_B \ ,$$
$$\partial n_B/\partial t = K_e C_B n_A - K_e C_A n_B \ . \qquad (2.56)$$

Sometimes, e.g., in masers, polarized spins take part in the exchange. In these cases it may be necessary to use the initial equations of the type (2.52) nonlinear with respect to ρ^A, ρ^B.

Spin exchange processes are usually studied by magnetic resonance methods. It is appropriate, therefore, to obtain equations for the changes in the magnetization $\underline{M}^A(\underline{M}^B)$ of particles. If the system in question is placed into a constant magnetic field $\underline{H}_0$ the direction of the latter is the axis of spin quantization. In this case directly from (2.56) we derive kinetic equations for the relaxation of the longitudinal magnetizations $M_z^A(M_z^B)$ of the spins of A(B), respectively. Indeed, the longitudinal magnetization of spins is proprotional to the difference of the concentrations of spins oriented in the direction of, and against, the field, i.e., $M_z^A \sim n_A$, $M_z^B \sim n_B$ [see (2.29)]. Therefore from (2.56) we obtain

$$\partial M_z^A/\partial t = -K_e C_B M_z^A + K_e C_A M_z^B \ ,$$
$$\partial M_z^B/\partial t = K_e C_B M_z^A - K_e C_A M_z^B \ . \qquad (2.57)$$

Thus, the flip-flop of spins brought about by the exchange interaction of paramagnetic particles A and B in the course of their binary collisions results in a reversible process of exchanging the longitudinal magnetizations with a rate constant K_e.

Now let us turn to the relaxation of the transverse components of magnetization. This magnetization is conveniently assigned by the quantity

$$M_- = M_x - iM_y \qquad (2.58)$$

which, according to (2.29), is proportional to a nondiagonal element $\rho_{\alpha\beta}$ of the spin density matrix. The variation of M_- is contributed to by a Larmor precession of $\underline{M}$ in the external magnetic field and by spin exchange processes. The alteration of M_- due to the Larmor precession is described by the well-known equation [2.22]

$$\partial M_-^A/\partial t = i\omega_A M_-^A \ ,$$
$$\partial M_-^B/\partial t = i\omega_B M_-^B \ . \qquad (2.59)$$

The exchange interaction in binary collisions of particles was shown above to cause a relaxation of the transverse components of magnetization (phase relaxation) through two mechanisms. If a collision of spins results in an equivalent change of their z projections, a simultaneous exchange of x and y projections must take place as well [see (2.16)]. Therefore, similar to (2.57) one can write the kinetic equations for the transverse components of magnetization

$$\partial M_-^A/\partial t = i\omega_A M_-^A - K_e C_B M_-^A + K_e C_A M_-^B ,$$

$$\partial M_-^B/\partial t = i\omega_B M_-^B + K_e C_B M_-^A - K_e C_A M_-^B \tag{2.60}$$

which contain the same rate constant K_e. Below we shall see that, as a rule, such an equivalent exchange of spin states turns out to be the main mechanism of phase relaxation in diluted solutions of organic free radicals.

The other mechanism of relaxing the tansverse components of magnetization is the dephasing of spins in colliding particles (see Fig.2.6). The alteration of the spin phase in a collision is similar to a bimolecular irreversible reaction.

Therefore the relaxation due to spin dephasing is described by the equations

$$\partial M_-^A/\partial t = -K_{eA}^{'} C_B M_-^A ,$$

$$\partial M_-^B/\partial t = -K_{eB}^{'} C_A M_-^B , \tag{2.61}$$

where $K_e^{'}$ stands for the rate constant of the dephasing process in binary collisions of particles. As will be seen below, when free radicals collide with paramagnetic complexes having short paramagnetic relaxation times the spin dephasing according to (2.61) constitutes the principal mechanism of relaxing the transverse components of magnetization.

In real systems both mechanisms of phase relaxation may be working simultaneously. Then the kinetic equations for the transverse components of magnetization take the form

$$\partial M_-^A/\partial t = (i\omega_A - K_e C_B - K_{eA}^{'} C_B)\, M_-^A + K_e C_A M_-^B ,$$

$$\partial M_-^B/\partial t = (i\omega_B - K_e C_A - K_{eB}^{'} C_A)\, M_-^B + K_e C_B M_-^A . \tag{2.62}$$

As shown by a more detailed analysis, exchange interaction may also produce a shift σ_e of resonance frequencies ω_A and ω_B [see(2.5)].

Kinetic Equations for Free Radicals Showing Hyperfine Structure (HFS) in Their ESR Spectra

Equations (2.62) can be generalized for more complicated situations [2.28]. By way of example let us consider a solution of free radicals with an arbitrary hfs. As a result of collisions of spins with an effective frequency K_eC an exchange of magnetization occurs between the spins. The decrease in the magnetization $\underline{M}(\omega_k)$ of the spins of the kth component characterized by a resonance frequency ω_k is described by the equation

$$\partial\underline{M}(\omega_k)/\partial t = -K_eC[1 - \varphi(k)]\,\underline{M}(\omega_k) \ , \tag{2.63}$$

where $\varphi(k)$ is the statistical weight of the spins in the kth component; K_eC is the rate of exchange of states between a given spin and all the others including those from this same component. Since the exchange between the spins of the same component does not change the fractional magnetization $M(\omega_k)$, a factor $1 - \varphi(k)$ appears in (2.63). This factor equals the probability of a collision between spin of the kth component and spins of all other components. The fractional magnetization of the singled out component increases due to the flux of magnetization from the other components as a result of the spin exchange which is described by the equation

$$\partial\underline{M}(\omega_k)/\partial t = K_eC\varphi(k)\sum_{n(\neq k)}\underline{M}(\omega_n) \quad . \tag{2.64}$$

Summarizing (2.63) and (2.64) and taking into account a possibile additive contribution to relaxation from the mechanism of spin dephasing in a collision and a shift of the resonance frequencies of spins, the kinetic equations for the fractional magnetizations of individual components may be represented in the form

$$\begin{aligned}\partial M_-(\omega_k)/\partial t &= i(\omega_k + \sigma_{ek})\,M_-(\omega_k) - K'_{ek}C[1-\varphi(k)]\,M_-(\omega_k)\\ &\quad -K_eC[M_-(\omega_k) - \varphi(k)\sum_n M_-(\omega_n)] \ ,\\ \partial M_z(\omega_k)/\partial t &= -K_eC[M_z(\omega_k) - \varphi(k)\sum_n M_z(\omega_n)] \quad .\end{aligned} \tag{2.65}$$

The kinetic equations obtained describe phenomenologically the temporal changes in the macroscopic magnetization of the system that are brought about by spin exchange processes in the course of binary collisions of paramagnetic species. They constitute the basis for the analysis of spectral manifestations of spin exchange.

It should be emphasized that apart from the spin exchange processes in real situations one must also take account of some other mechanisms of paramagnetic relaxation.

2.4 Spectroscopic Manifestations of Spin Exchange

According to the discussion in Sect.2.2 the exchange interaction of paramagnetic particles in solutions may lead to two qualitatively different results: spin dephasing or equivalent exchange of spin states. In real systems the equivalent exchange is frequently predominant.

The spectroscopic manifestation of a dipole moment dephasing is well known. The relaxation mechanism arising from a dephasing of spin precession during binary collisions reduces the time T_2 of transverse paramagnetic relaxation of particles. As a result, the spin dephasing due to exchange interaction leads to an additional broadening of ESR lines of paramagnetic particles.

A relaxation mechanism brought about by an equivalent spin exchange spectroscopically manifests itself in a more complex manner. Therefore we shall dwell in more detail on the analysis of this practically most important mechanism.

Spectral Diffusion Due to Spin Exchange

Of great importance is the fact that formally spectroscopic manifestations of spin exchange can be analyzed in terms of a model of frequency migration or of spectral diffusion [2.28-34]. The fundamental study of the spectral manifestations of frequency migration including those induced by exchange interaction was carried out in the well-known pioneering works by WEISS and ANDERSON [2.29,30].

Their results can be made use of in analyzing the spectral manifestations of spin exchange processes.

Indeed, (2.57,60) describing the dynamic behavior of magnetization in the presence of spin exchange between two types of paramagnetic particles coincide with the equations describing this behavior for chemical exchange of a particle between two sites with different Larmor frequencies equal to ω_A in one site and to ω_B in the other [2.35]. In a more general case described by (2.65) the kinetic equations for fractional magnetizations also coincide with the corresponding kinetic equations in a model of frequency migration among many values [2.28-34].

In the model of frequency migration it is assumed that the Larmor frequency of a particle is changed randomly. Besides, the frequency change (jump) itself is assumed to occur so quickly that during the jump the magnetization vector fails to change its orientation. Within the framework of this model the axis of spin precession is retained. Therefore the motion of say transverse magnetization is described by the stochastic equation

$$\partial M_-/\partial t = i\omega(t)M_- \quad . \tag{2.66}$$

Random changes of $\omega(t)$ can be modelled by the Markov processes [2.36]. The Markovian process $\omega(t)$ can be assigned by conditional transition probabilities. Let $p(\omega,t|\omega',t')$ designate the conditional probability that ω has the value ω at a time t given that at a previous moment t' its value was ω'. If in (2.66) $\omega(t)$ represents a stationary Markovian process with the conditional probability

$$\begin{aligned} p(\omega,t|\omega',t') &= \exp -K_eC(t-t')\ \sigma(\omega-\omega') \\ &\quad + \{1 - \exp[-K_eC(t-t')]\}\ \varphi(\omega) \quad , \end{aligned} \tag{2.67}$$

where σ is a delta function, and $\varphi(\omega)$ is the density of states with a frequency ω, then (2.66) can be shown [2.32,37] to be really equivalent to the equations of (2.65) for the fractional magnetizations $M_-(\omega_k)$.

Shape of ESR Spectra

The spin exchange process manifests itself in rather characteristic transformations of the ESR spectra as the rate of exchange increases. Depending on the relation between the exchange rate equal to $\omega_e = K_eC$ and the difference of resonance frequencies of spins belonging to different components of the ESR spectrum the exchange causes qualitatively different changes in the spectrum. The spread of spin frequencies can be characterized by their dispersion equal to $\langle\Delta\omega^2\rangle$. The case (2.68),

$$\omega_e < \langle\Delta\omega^2\rangle^{\frac{1}{2}} \quad , \tag{2.68}$$

is called the situation of slow exchange: under this condition spin exchange broadens individual components of the spectrum, and shifts the position of their maxima. The opposite case

$$\omega_e > \langle\Delta\omega^2\rangle^{\frac{1}{2}} \tag{2.69}$$

is referred to as the situation of fast exchange. In this case an effect of exchange narrowing is observed: all the components of the ESR spectrum merge into one narrow line at the center of the spectrum.

These phenomena will now be discussed in more detail. In ESR experiments we measure the absorption of the power of the microwave field H_1 which rotates at an angular frequency ω in a plane perpendicular to $\underline{H}_0$. This absorption is due to the fact that the magnetization of a sample fails to instantly follow the changes of H_1, i.e., falls out of phase with the microwave field. In a coordinate system rotating about the z direction ($z \parallel \underline{H}_0$) with a frequency ω and with $\underline{H}_1$ vector lying along the x axis the measured absorption power is expressed through the steady-state value of the magnetization projection on the y axis, $\tilde{M}_y(\omega)$, by means of the equations [2.31]

$$dA/dt = -(\underline{M} \cdot d\underline{H}_1/dt) = -\omega H_1 \tilde{M}_y(\omega) \quad . \tag{2.70}$$

From (2.70) the frequency dependence of absorption is seen to be determined by two factors: ω and $\tilde{M}_y(\omega)$. However, the relative alterations of ω over the absorption band are negligible, as a rule. Therefore the frequency dependence of the power losses or the shape $g(\omega)$ of the spectra is, in fact, determined only by $\tilde{M}_y(\omega)$, i.e., by one of the following equivalent equations

$$g(\omega) \sim \tilde{M}_y(\omega) \ ,$$
$$g(\omega) \sim \mathrm{Im}\{M_-(\omega)\} \quad . \tag{2.71}$$

To find the steady-state value of the magnetization of the sample, we write a kinetic equation for fractional magnetizations of free radicals belonging to different hf components of the ESR spectrum

$$\begin{aligned} d\tilde{M}_-(\omega_k)/dt = {} & [i(\omega_k - \omega) - 1/T_{2k}]\, \tilde{M}_-(\omega_k) \\ & -K_e C[\tilde{M}_-(\omega_k) - \varphi(k) \sum_n \tilde{M}_-(\omega_n)] - i\omega_1 M_z(\omega_k) \ , \end{aligned} \tag{2.72}$$

$$\begin{aligned} dM_z(\omega_k)/dt = {} & -\omega_1 \tilde{M}_y(\omega_k) - [M_z(\omega_k) - M_0(\omega_k)]/T_{1k} \\ & - K_e C[M_z(\omega_k) - \varphi(k) \sum_n M_z(\omega_n)] \ , \end{aligned}$$

where $\omega_1 = \gamma H_1$; γ is a gyromagnetic ratio for the electron; H_1 is the amplitude of the microwave field, $M_0(\omega_k)$ is the equilibrium magnetization of the component. These equations are derived from (2.65), if they are written in the rotating coordinate system and the field H_1 is added.

A detailed consideration of paramagnetic relaxation mechanisms not associated with the process of spin exchange is beyond the scope of this book.

We shall therefore take them into account by introducing some phenomenological parameters which in (2.72) are the times of the longitudinal (spin-lattice) T_{1k} and transverse (phase) T_{2k} relaxation of spins belonging to the kth component. The times T_{1k} and T_{2k} do not include the effect of an equivalent spin exchange. To simplify the further calculations consider a case when all components are characterized by equal relaxation times T_1, T_2.

Between the microwave field and the system of spins a reversible transfer of energy occurs. Denote the rate of this process by W_s. The energy absorbed by spin is transferred, at a velocity $1/T_1$ into the lattice, i.e., is turned into heat. When $W_s < 1/T_1$, i.e., in sufficiently weak fields $H_1(t)$, the spin-lattice relaxation ensures a sufficiently fast transformation of the absorbed energy into heat. In this case the total energy of the spin system does not practically change when the field $H_1(t)$ is switched on. The latter does not heat up the spin system which is in good contact with the lattice, which acts as a thermostat. In the other limiting case [strong fields $H_1(t)$] when $W_s > 1/T_1$ the steady-state absorption is already limited by the transfer of spin energy into the lattice. When the field $H_1(t)$ is switched on the full energy of the spin system may markedly increase and inhibit any further absorption. The well-known phenomenon of spin system saturation occurs.

Let us first consider the form of the spectra in the weak fields H_1. Then we can neglect the change of longitudinal magnetization; hence [2.31]

$$M_z(\omega_k) = M_0(\omega_k) \cong M_0\varphi(k) \quad . \tag{2.73}$$

Substituting (2.73) into (2.72) we find that the steady-state value of M_- and, as a result, the shape of the ESR spectrum are determined by the equations

$$\{i(\omega_k-\omega) - 1/T_2 - K_eC[1-\varphi(k)]\}\,\tilde{M}_-(\omega_k) + K_eC\varphi(k) \sum_{n(=k)} \tilde{M}_-(\omega_n) = i\omega_1 M_0\varphi(k) \equiv w_k \quad . \tag{2.74}$$

From (2.74) the equation for the total magnetization of the system follows directly,

$$\tilde{M}_- = \sum_n \tilde{M}_-(\omega_n) = i\omega_1 M_0 G/(1 + K_eCG) \ , \tag{2.75}$$

here

$$G = \sum_n \varphi(n)/[i(\omega_n-\omega) - 1/T_2 - K_eC] \quad .$$

From (2.70) and (2.75) we find that the power absorbed by the spin system is equal to

$$dA/dt = \omega\omega_1^2 M_0 \, \mathrm{Im}\{iG/(1 + K_e CG)\} \quad . \tag{2.76}$$

The shape of the ESR spectrum is given by the equation

$$g(\omega) \sim \mathrm{Im}\{iG/(1 + K_e CG)\} \quad . \tag{2.77}$$

In the case of multicomponent spectra the analysis of the general expression (2.77) is a rather cumbersome task. Relatively simple results are obtained only in the limiting cases of slow and fast exchange.

Slow Exchange

In the case of slow exchange equations (2.74) can be solved using the theory of perturbation with respect to the small parameter $\varepsilon = \omega_e / <\Delta\omega^2>^{\frac{1}{2}}$ for the nondegenerate case [2.6]. In the zero approximation

$$\tilde{M}_-(\omega_k) = i\omega_1 M_0 \varphi(k)/\{i(\omega_k - \omega) - 1/T_2 - K_e C[1-\varphi(k)]$$

and the ESR spectrum represents a superposition of Lorentzian curves

$$g(\omega) = \sum \varphi(k)\Delta\Omega_k/\pi[(\omega-\omega_k)^2 + \Delta\Omega_k^2] \quad . \tag{2.78}$$

The halfwidth of each hf component at the half-height is equal to

$$\Delta\Omega_k = 1/T_2 + K_e C[1 - \varphi(k)] \quad . \tag{2.79}$$

Thus in the situation under study the spin exchange produces an additional broadening of the lines by the quantities

$$\Delta\Omega_{ek} = K_e C[1 - \varphi(k)] \quad . \tag{2.80}$$

Also, the spin exchange shifts the position of the maxima of the ESR spectrum lines to the center of the spectrum. The shift of the maxima of the lines is due to two reasons: first the exchange varies the characteristic roots of (2.74) and, second, it distorts the line shapes of individual components of the spectrum. Let ω_{ek} be the roots of a determinant made up of the coefficients of the left-hand side of (2.74). With their help the solution of (2.74) may be presented as the sum

$$\tilde{M}_- \sim \sum_k a_k/(\omega - \omega_{ek}) \ ,$$

where the coefficients a_k are independent of the frequency ω, but in the general case they are complex as are the characteristic roots $\omega_{ek} : a_k = u_k + iv_k$, $\omega_{ek} = \omega'_k + i\Gamma_k$. Then the line shape of an individual k^{th} component of ESR spectrum is described by

$$g_k(\omega) \sim \{\mathrm{Im}\{a_k/(\omega-\omega_{ek})\} = [u_k\Gamma_k + v_k(\omega-\omega_k')]/[(\omega-\omega_k')^2 + \Gamma_k{}^2] \quad . \tag{2.81}$$

Hence the line shape, as distinct from the result of the zero approximation (2.78) is no longer described by the Lorentzian curve. The line becomes asymmetric with respect to ω_k'. In the region of slow exchange when the asymmetry parameter v_k is a quantitiy of the order of $\varepsilon < 1$, the maximum of (2.81 is attained at the frequency

$$\omega_{k,max} \cong \omega_k' + \Gamma_k v_k/2u_k \quad . \tag{2.82}$$

The position of the line maximum shifts by amount $\omega_k' - \omega_k$ as a result of the variation of the characteristic roots of (2.74), and by the amount $\Gamma_k v_k/2u_k$ due to the line shape distortion, either of which is caused by the spin exchange. Let us find both of these contributions of the exchange to the shift of lines.

In second-order perturbation theory the characteristic roots for system (2.74) are equal to

$$\omega_{ek} = \omega_k' + i\Gamma_k = \omega_k + K_e^2C^2\varphi(k) \sum_{n(\neq k)} \varphi(n)/(\omega_n-\omega_k)$$
$$+i\{1/T_2 + K_eC[1-\varphi(k)]\} \quad . \tag{2.83}$$

From (2.82) and (2.83) it follows that the exchange produces a line shift which is proportional to the square of the concentration of paramagnetic particles

$$\sigma_{ek}' = \omega_k' - \omega_k = K_e^2C^2\varphi(k) \sum_{n(\neq k)} \varphi(n)/(\omega_n-\omega_k) \quad . \tag{2.84}$$

Before trying to find the shift of lines due to the distortion of their shape note the following. The quantity v_k expands into a series in odd powers of the parameter ε while u_k, in even ones. From (2.82) it can be seen that to find the shift correct to second order in ε it is sufficient to obtain u_k in the zero, and v_k in the first order. Up to the first order in the small parameter ε the components of the mangetization vector and the right-hand side of (2.74) take the form

$$\tilde{M}_-'(\omega_k) = \tilde{M}_-(\omega_k) - iK_eC\varphi(k) \sum_{n(\neq k)} \tilde{M}_-(\omega_n)/(\omega_k-\omega_n) \ ,$$
$$w_k' = i\omega_1M_0\varphi(k)[1 - iK_eC \sum_{n(\neq k)} \varphi(n)/(\omega_k-\omega_n)] \tag{2.85}$$

and (2.74) is reduced to the system of uncoupled equations

$$(\omega-\omega_{ek})\tilde{M}'_-(\omega_k) = iw'_k \quad .$$

Solving the latter equations and taking into account that in first approximation

$$\tilde{M}_-(\omega_k) = \tilde{M}'_-(\omega_k) + iK_eC\varphi(k) \sum_{n(\neq k)} \tilde{M}'_-(\omega_n)/(\omega_k-\omega_n)$$

we obtain the total magnetization of the system

$$\tilde{M}_- = \sum_k \tilde{M}_-(\omega_k) = \sum_k (u_k+iv_k)/(\omega-\omega_{ek}) \quad .$$

As a result we obtain

$$u_k \cong -\omega_1 M_0 \psi(k) \ ,$$
$$v_k \cong \omega_1 M_0 K_e C \sum_{n(\neq k)} 2\varphi(k)\varphi(n)/(\omega_k-\omega_n) \quad .$$

Substituting these results into (2.82) we obtain the magnitude of the shift of the spectrum line maxima due to their non-Lorentzian form

$$\delta''_{ek} = \{1/T_2 + K_eC[1-\varphi(k)]\}K_eC \sum_{n(\neq k)} \varphi(n)/(\omega_n-\omega_k) \quad . \tag{2.86}$$

The total shift of the line maximum is equal to

$$\delta_{ek} = (1/T_2 + K_eC)K_eC \sum_{n(\neq k)} \varphi(n)/(\omega_n-\omega_k) \quad . \tag{2.87}$$

If the line width in the absence of exchange is greater than the exchange broadening (2.77), then the distortion of the line shape makes the main contribution to the shift of the absorption line maximum. This shift increases linearly with the concentration of paramagnetic species

$$\delta_{ek} \cong T_2^{-1}K_eC \sum_{n(\neq k)} \varphi(n)/(\omega_n-\omega_k) \quad . \tag{2.88}$$

When spin exchange becomes a dominant broadening mechanism, i.e., $K_eC > 1/T_2$, the line is shifted quadratically in concentration of particles. In this case the line shift is determined by both of the above mechanisms and their total effect gives the shift

$$\delta_{ek} \cong K_e^2C^2 \sum_{n(\neq k)} \varphi(n)/(\omega_n-\omega_k) \quad . \tag{2.89}$$

For a two-component spectrum when $\varphi(1) = \varphi(2) = 1/2$ the contributions of the two mechanisms to the shift are equal to each other. As the number of the spectral components increases the relative contribution to the line shift due to the distortion of the line shape increases in the ratio $[1-\varphi(k)] : \varphi(k)$ [compare (2.84) with (2.86)].

Fast Exchange

Now we turn to the discussion of ESR spectrum in the situation of fast exchange. To find the shape of the spectrum in this case one may solve (2.74) as before and analyze the result obtained under the fast exchange condition $\omega_e > \langle\Delta\omega^2\rangle^{1/2}$ [2.28,38]. It is more convenient, however, to analyze the spectrum shape with the help of the correlation theory of linear response [2.31]. According to this theory the shape of the absorption spectrum is determined by the Fourier transform of the relaxation function of the transverse component of the system mangetization

$$g(\omega) \sim \mathrm{Re}\left\{\int_0^\infty dt \langle M_-(t)M_+(0)\rangle \exp(i\omega t)\right\} \tag{2.90}$$

where $\langle\ldots\rangle$ stands for the averaging over the ensemble of systems. The time variation of magnetization can be found from (2.66)

$$M_-(t) = M_-(0) \exp\left[i \int_0^t \omega(t')dt' - t/T_2\right] \quad . \tag{2.91}$$

Substituting (2.91) into (2.90) we obtain

$$g(\omega) \sim \mathrm{Re}\left\{\int_0^\infty dt \langle \exp[i \int_0^t \omega(t')dt'] \rangle \exp(i\omega t - t/T_2)\right\} \quad . \tag{2.92}$$

Denoting the mean frequency by ω_0 we have $\omega(t) = \omega_0 + \Delta\omega(t)$. The magnetization relaxation function in the case of fast exchange can be obtained by using the stochastic perturbation theory [2.31]. Having this in mind we write it as the series

$$\langle \exp[i \int_0^t \Delta\omega(t')dt'] \rangle = 1 - (1/2) \int_0^t\int_0^t dt'dt'' \langle \Delta\omega(t')\Delta\omega(t'')\rangle + \ldots \quad . \tag{2.93}$$

Under the fast exchange conditions the correlation time of frequency migration $\tau_e = 1/K_eC$ is relatively short, i.e., $\tau_e^2 \langle\Delta\omega^2\rangle < 1$. The relaxation of magnetization occurs within the times t that are much longer than τ_e. Therefore in (2.93) it is necessary to consider the times $t \gg \tau_e$. When these conditions are satisfied the omitted terms of series (2.93) prove to be very small quantities, of the order of $\tau_e^2 \langle\Delta\omega^2\rangle$ or less, and can be neglected [2.31]. Using the conditional probability (2.67) we obtain

$$\langle\Delta\omega(t')\Delta\omega(t'')\rangle = \iint d\omega'\cdot d\omega''\varphi(\Delta\omega') \times p(\Delta\omega'', t''|\Delta\omega', t')\Delta\omega'\cdot\Delta\omega''$$
$$= \langle\Delta\omega^2\rangle \exp(-K_eC|t'-t''|) \quad .$$

Substituting this expression into (2.92) we have

$$\langle \exp[i \int_0^t \Delta\omega(t')dt'] \geqq \exp(-\langle\Delta\omega^2\rangle t/K_eC) \quad . \tag{2.94}$$

Finally we obtain the exchange-narrowed absorption curve which was obtained earlier by ANDERSON and WEISS [2.29]

$$g(\omega) = \Delta\Omega/\pi[(\omega-\omega_0)^2 + \Delta\Omega^2] \ , \tag{2.95}$$

where

$$\Delta\Omega = \langle\Delta\Omega^2\rangle/K_eC + 1/T_2 \quad .$$

As a result of the fast exchange all the spectrum components merge into one Lorentzian line with the maximum in the center of the spectrum whose width decreases as the exchange rate increases,

$$\Delta\omega_e = \langle\Delta\omega^2\rangle/K_eC \quad . \tag{2.96}$$

This phenomenon is referred to as exchange narrowing of absorption spectra.

Thus, the rate of equivalent spin exchange can be found from experimental data on the concentration broadening and on the shift of spectrum components or from the line width in the case of exchange narrowing of the spectrum.

Other Methods of Measuring the Rate of Spin Exchange

Most of the data available in literature on spin exchange between paramagnetic species have been obtained by analyzing the spectra in the absence of saturation. Therefore we have so far focused our attention on the spin exchange manifestations in the shape of nonsaturated ESR spectra. This method is attractive due to the simplicity and feasibility of the experiment and the reliability of the theoretical interpretation of the results.

However the possibilities of this method are limited. For example, along with the process of equivalent spin exchange, spin dephasing due to relatively weak exchange interaction in binary collisions, and dipole-dipole spin-spin interactions may contribute to the concentration broadening of the absorption lines. Thus the problem arises of isolating the contribution from the equivalent exchange. Besides, there may be difficulties in measuring the exchange broadening and the line shift in the case of poorly resolved multicomponent spectra. For some systems the exchange narrowing of the spectrum may prove unattainable due to the impossibility of obtaining a sufficiently high concentration of paramagnetic species as a result of their limited solubility, recombination in the case of triplet excitons and free radicals, etc. To overcome these difficulties one has to resort to methods based on other spectral effects of spin exchange.

The shape of nonsaturated ESR spectra contains only data on transverse relaxation. In the general case the exchange interaction during binary collisions of spin bearing molecules may change transverse and longitudinal spin projections with different efficiency. Only in the case of equivalent spin exchange are they changed at the same rate. Therefore those methods depending on the exchange of longitudinal magnetization between the components may yield independent information about the rate of equivalent spin exchange. The Zeeman energy E(k) of the spins of a given spectrum component is proportional to their longitudinal magnetization: $E(k) \sim M_z(\omega_k)$. The flip-flop of spins that belong to different components induces a transfer of Zeeman energy between them. According to (2.72) the kinetics of redistributing the Zeeman energy due to spin exchange is described by

$$\partial E(k)/\partial t = -K_eC[E(k) - \varphi(k) \sum_n E(n)] \quad . \qquad (2.97)$$

In order to pump energy into the spin system one must irradiate the sample with a microwave field of a sufficiently large amplitude, so the measurement of the rate of energy redistribution in the spin system involves experiments with strong microwave fields.

Effect of Saturation

As the power of the microwave field rises to sufficiently large values the spin system is saturated and the energy transfer from the spin system to the lattice, a heat reservoir, becomes a limiting stage of steady-state absorption. The redistribution of energy within the spin system may greatly affect the saturation. The shape of ESR spectra in the case of saturation can be established by solving (2.72) for the steady-state conditions, i.e., assuming $d\underline{M}(\omega_k)/dt = 0$. The equations obtained are formally solved rather easily. For this purpose we rewrite them in the form

$$-(1/T_2 + K_eC)\tilde{M}_x(\omega_k) + (\omega_k-\omega)\tilde{M}_y(\omega_k) = -\varphi(k)K_eC\tilde{M}_x \ ;$$
$$-(\omega_k-\omega)\tilde{M}_x(\omega_k) - (1/T_2 + K_eC)\tilde{M}_y(\omega_k) + \omega_1M_z(\omega_k) = -\varphi(k)K_eC\tilde{M}_y \ ;$$

$$-\omega_1\tilde{M}_y(\omega_k) - (1/T_1 + K_eC)M_z(\omega_k) = -\varphi(k)(K_eCM_z + M_0/T_1) \ , \qquad (2.98)$$

where

$$\tilde{M}_{x,y} \equiv \sum_k \tilde{M}_{x,y}(\omega_k) \ , \quad M_z \equiv \sum M_z(\omega_k) \quad .$$

From this system of algebraic equations the fractional magnetization $\tilde{M}_x(\omega_k)$, $\tilde{M}_y(\omega_k)$, $M_z(\omega_k)$ can be expressed in terms of the total magnetization $\tilde{M}_x, \tilde{M}_y$, M_z and the equilibrium magnetic moment M_0 of the sample. The solution of (2.98) has the form

$$\underline{\tilde{M}}(\omega_k) = A_k\underline{\tilde{M}} + B_k\underline{M}_0 \ , \qquad (2.99)$$

where the matrices A_k and B_k are determined from (2.98) by standard methods of linear algebra [2.39]. Summing up (2.99) we obtain

$$\underline{\tilde{M}} = (\sum_k A_k)\underline{M} + (\sum_k B_k)\underline{M}_0 \quad . \qquad (2.100)$$

For $\tilde{M}_x$, $\tilde{M}_y$, M_z we have a system of three algebraic equations. Although in principle the method of finding the total magnetization of a system is quite clear, the explicit form of solutions for particular systems is rather complicated. It should be noted that the effects of saturation in a particular case of exchange between two positions have been discussed in [2.35]. In [2.40] a general solution of (2.98) has been obtained for an exchange between many frequencies on the assumption that $T_1 = T_2$, and an analysis has been made of the saturation for the Gaussian distribution $\varphi(\omega_k)$. The saturation of the Gaussian and Lorentzian lines at arbitrary T_1 and T_2 has been studied in [2.41,42], respectively. A general theoretical analysis of the saturation effects of the ESR spectra in the presence of spin exchange was made by FREED et al. [2.43-45].

The study of saturation effects in the case of fast exchange when the whole spectrum constitutes a single homogeneously broadened line does not yield any new data on spin exchange as compared with the analysis of the line shape in nonsaturating fields. The saturation is described by the T_1- T_2 model [2.46-48]

$$dA/dt = \omega\omega_1^2M_0[1 + \omega_1^2T_1T_2]^{-\frac{1}{2}} \cdot T_2' \times [1 + (\omega-\omega_0)^2T_2'^2]^{-1} \ , \qquad (2.101)$$

where ω_0 is the mean frequency, $1/T_2 = \Delta\Omega$ is the line width equal to (2.95), and

$$T_2' = T_2/(1 + \omega_1^2T_1T_2)^{\frac{1}{2}} \quad . \qquad (2.102)$$

For slow exchange, when $K_eC < \langle\Delta\omega^2\rangle^{\frac{1}{2}}$ (2.68), the absorption curve can be presented as a superposition of independently saturated lines, each of these being described by the $T_1 - T_2$ model. The effect of spin exchange is taken into account by introducing the effective parameter $T_1(k)$ and $T_2(k)$ for each component. For multicomponent spectra with a resolved structure under the slow exchange conditions the time of transverse relaxation of individual components is given by (2.79). As to the effective time of longitudinal relaxation of individual components, it is influenced by spin exchange differently depending on the relation between the rate of exchange and that of spin-lattice relaxation. The microwave energy is directly absorbed only by the spins that are resonant with it. If

$$K_eC < 1/T_1 \tag{2.103}$$

then the nonresonance spins can be considered as a part of the lattice since the return of excitation from the nonresonance spins back to the resonance ones is unlikely in this condition. The decrease in the effective time of longitudinal relaxation due to the excitation transfer from the resonance spins to the others, unexcited by pumping, is described by the expression [cf (2.79)]

$$1/T_1(k) \simeq 1/T_1 + K_eC(1 - \varphi(k)) \quad . \tag{2.104}$$

If the redistribution of energy within a spin system proceeds faster than the transfer of spin excitation to the lattice, i.e.,

$$K_eC > 1/T_1 \ , \tag{2.105}$$

then all the spins, and not only the resonance ones, are saturated. The whole spin system absorbs the energy and is heated up as a single exchange-bound reservoir. The fast redistribution of the excitation throughout the spin system increases the rate of the excitation energy transfer to the lattice since the process involves all the spins of the system, not only the resonance ones. The rate of spin-lattice relaxation of the spins of a given hf component increases as many times as the heat capacity c_0 of the entire system exceeds the heat capacity c_k of the spins of the component in question

$$T_1(k) = T_1c_k/c_0 = T_1\varphi(k) \quad . \tag{2.106}$$

The general formula for the effective longitudinal relaxation time at an arbitrary relation between K_eC and T_1 in the case of slow exchange was obtained by FREED et al. in [2.44]

$$T_1(k) = T_1[1 + \varphi(k)K_eCT_1] \cdot (1 + K_eCT_1)^{-1} \quad . \tag{2.107}$$

Expressions (2.104) and (2.106) are the extreme forms of (2.107).

With the help of the effective parameters $T_1(k)$ and $T_2(k)$ the saturation of the spectrum in the region of slow exchange can be described using the theory of steady-state saturation of inhomogeneously broadened spectra with no allowance for spectral diffusion [2.46,47]. We obtain

$$dA/dt \cong \omega\omega_1^2 M_0 \sum \varphi(k)T_2'(k) \times [1 + \omega_1^2 T_1(k)T_2(k)]^{-\frac{1}{2}} \cdot [1 + (\omega_{ek}-\omega)^2 T_2'^2(k)]^{-1} \quad , \tag{2.108}$$

where $T_2'(k)$ is connected with $T_2(k)$ by (2.102). As a result of a decrease in the effective time of longitudinal relaxation the saturation of an individual component of the spectrum may occur at far greater values of the amplitude of the pumping field than might be the case in the absence of excitation transfer in a spin system.

Thus, by analyzing the data on steady-state sturation one can determine the rate of energy redistribution in a spin system. The main advantage of the method of steady-state saturation over the methods based on measuring an exchange broadening or a shift of lines is that it allows the exchange rate to be measured at much lower concentrations of paramagnetic species. Indeed, as a rule, $T_2 < T_1$. Therefore one can detect the influence of spin exchange on T_1 at lower concentrations.

Direct measurements of the rate of spin excitation transfer between spectrum components can be made by various two-frequency methods in a pulse and in steady-state regimes.

Electron-Electron Double Resonance (ELDOR)

There are several ways of conducting experiments on ELDOR. An ELDOR experiment consists of irradiating one part of the ESR spectrum with a microwave source and monitoring the resulting response at another portion of the spectrum with a second microwave source [2.49,50]. The ELDOR technique was first used by HYDE [2.49] for the study of spin exchange in liquids.

In the pulse ELDOR technique the recovery of the signal intensity is recorded after the pumping is switched off [2.49,51]. If $K_eC > 1/T_1$, then the pumping field saturates uniformly all the spins. The recovery of all spectrum components will proceed at the same rate $1/T_1$. In this case the kinetics of the spectrum recovery yields the lower estimate of the rate of the spin excitation transfer. If $K_eC \lesssim 1/T_1$, then the pumping saturates the resonance spins to a greater extent than the nonresonance ones. Thus the pump succeeds in burning a hole in the spectrum. The filling in of this hole occurs due to a redistribution of energy among all the spins, on the one hand, and as a result of a spin-lattice relaxation, on the other. In this case an analysis of the subsequent return of the signal to equilibrium enables one to determine the rate of the spin excitation transfer [2.52].

Consider a case when there is an exchange between two lines of equal intensity. If by the moment of the switch-off of the pumping the values of the longitudinal magnetizations of the components were equal to $M_z^A(0)$ and $M_z^B(0)$, then the subsequent return to the equilibrium spectrum will proceed as follows: at the rate $K_eC + 1/T_1$ the intensity of both components will be equalized and at the rate $1/T_1$ the whole spectrum will be recovered to its equilibrium value. Therefore, by monitoring the kinetics of the signal intensity recovery one can obtain the rate of energy redistribution in a spin system. The above method of investigating excitation transfer in a spin system is widely applied in the study of solids [2.52-54]. However, as has been pointed out by HUISJEN and HYDE [2.51], the application of the double resonance technique to the study of relaxation processes in a liquid may present some difficulties. The fact is that just after the pumping field is switched off the system possesses a nonzero transverse magnetization which gives a free induction signal. In a liquid this signal decays within a time comparable with that of hole filling-in. In the experiment, therefore, special measures have to be taken to isolate the signal of free induction from a total monitoring field signal. For the separation, use may be made, e.g., of the fact that the free induction signal is independent of the monitoring field power, or one may change the phase of the monitoring field with respect to that of the pumping field [2.51]. Besides, one can increase the rate of the free induction signal decay by artificial devices, say, by increasing the inhomogeneity of the external magnetic field H_0.

In the spin exchange studies more often the steady-state version of ELDOR is used [2.49,50]. It consists of recording the relative alteration of the monitor signal intensity in response to pumping at another frequency

$$b = 1 - I/I_0 , \tag{2.109}$$

where I_0 and I are the amplitudes of the ESR signal without and with pumping, respectively.

In the simplest case of exchange between two frequencies, say, for radicals whose ESR spectrum consists of two components A and B, the analysis of the

ELDOR effect is given in [2.55,56]. The pumping field is assumed to saturate the component A so that its longitudinal magnetization $M_z^A \to 0$. The ELDOR effect is observed on the component B. The steady-state value of its longitudinal magnetization can be found by solving the equation [see (2.72)]

$$dM_z^B/dt = -K_e C_A M_z^B + K_e C_B M_z^A - (M_z^B - M_0^B)/T_{1B} = 0 \quad .$$

In the case when the component A is saturated we have

$$M_z^B = M_0^B/(1 + K_e C_A T_{1B}) \quad .$$

Taking into account that the intensity of the ELDOR line is proportional to M_z^B we obtain the result

$$b = K_e C_A T_{1B}/(1 + K_e C_A T_{1B}) \quad . \qquad (2.110)$$

Just as was expected, b grows from 0 to 1 with increase in the rate of exchange. Indeed, the spectrum components are assumed to be sufficiently separated so that the direct transfer of the energy of the pumping to the component monitored is negligible. As a result of a spin exchange the excitation is transferred to the other component, so the observed signal intensity drops off.

Electron Spin Echo

Spin exchange causes the decay of electron spin echo signals. The manifestations of spin exchange in the kinetics of the echo signal decay are discussed in detail in [2.31]. The echo signal is a response of the spin system to a sequence of two (primary echo) or three (stimulated echo) short microwave pulses. In the situation of slow exchange the primary echo signal decay depends on the time interval τ between two pulses forming an echo according to

$$v(2\tau) \sim \sum_k \varphi(k) \cdot \exp(-2\tau\{1/T_2 + K_e C[1-\varphi(k)]\}) \quad . \qquad (2.111)$$

Especially convenient for measuring the rate of slow spin exchange is a stimulated echo. According to [2.34,57] the dependence of the stimulated echo signal on the time intervals between the first and the second (τ), the second and the third (T) pulses is described by the expression

$$v(2\tau + T) \approx \exp(-K_e CT - T/T_1) \cdot v(2\tau) \quad . \qquad (2.112)$$

In the case of fast spin exchange the contribution of this process to echo signal decay is given by

$$v(2\tau) \sim v(2\tau + T) \sim \exp(-2\tau \langle \Delta\omega^2 \rangle / K_e C) \quad . \tag{2.113}$$

Thus, at sufficiently large rates of spin exchange the echo signal decay ceases to depend on this process but continues to occur under the action of relaxation mechanisms unrelated to the exchange. This provides a way of separating the contributions of the exchange and dipole-dipole spin interactions to the paramagnetic relaxation [2.34].

The spectral manifestations of the spin exchange process are far from exhausted by those briefly considered above. There are still other manifestations. For example, the flip-flop of electron spins produces random changes of the electron-nucleus interaction. If this mechanism ensures a quicker modulation of the electron-nucleus interaction than the modulation due to the rotation of paramagnetic particles, then from an analysis of their NMR spectra information on the spin exchange between them [2.58] can be obtained. As another example the use of masers for spin exchange studies in gases should be mentioned. For instance, in [2.59,60] the spin exchange between hydrogen atoms and their exchange with other paramagnetic species were studied by using a hydrogen quantum generator. The phase relaxation rate of hydrogen atoms was calculated from data on the free induction signal decay of polarized hydrogen atoms. Spin exchange in gases is not the subject of this book. Note, however, that the possible manifestations of spin exchange in masers are quite similar to those in ESR spectra.

2.5 Theory of Sudden Collisions

The theory of spin exchange described in the two previous sections is phenomenological. Now, proceeding from the microscopic equations of spin motion we shall, firstly, derive strictly the macroscopic kinetic equations which in Sect. 2.3 were postulated on the basis of an analogy between spin exchange and other bimolecular processes, and, secondly, find explicit expressions for the spin exchange rate through the parameters of exchange interaction and those of molecular dynamics of paramagnetic species.

2.5.1 Model of Sudden Collisions

Basic Assumptions

In the theory of spin exchange processes wide use is made of the model of sudden collisions suggested by KIVELSON [2.61]. Keeping in mind that the exchange integral drops off very rapidly with increasing distance between paramagnetic molecules it is assumed that it is only at the moment of an impact when two paramagnetic molecules are next to each other that an interaction with a value of the exchange integral J suddenly occurs between them. The interaction just as suddenly vanishes after a time t_c when the particles separate. Thus, the change of the exchange integral with time during an encounter of two paramagnetic particles is assigned in the form

$$J(t) = \begin{cases} J & \text{when } t_0 \le t \le t_0 + t_c \; , \\ 0 & \text{when } t < t_0 \; , \quad t > t_0 + t_c \; . \end{cases} \qquad (2.114)$$

Here t_0 is the moment of coming into contact.

The approximation under discussion will yield satisfactory results if the switch-on and switch-off times of exchange interaction are sufficiently short. Assume the diffusion of particles occurs as a result of their random jumps from one position to another. Denote by t' the time during which the particle changes its position (jumps) and by δ the splitting of the spin states of a pair of colliding particles. The approximation of sudden collisions is applicable when the condition [2.62]

$$\delta \cdot t' < 1$$

is satisfied. In spin exchange this condition is in most cases satisfied. Indeed, t' is a time within which particles at the thermal velocity of $\sim 10^5$ cm/s cover a distance of $\sim 10^{-8}$ cm, i.e., $t' \sim 10^{-13}$ s. The value of δ characterizes the splitting of states as a result of the hyperfine interaction (hfi) with magnetic nuclei or the zero field splitting, i.e., $\delta \sim 10^8 - 10^{10}$ rad/s. The result is that $\delta \cdot t' \lesssim 10^{-3}$.

In the model of sudden collisions the escape of paramagnetic particles from the "cage" is regarded as a reaction of the monomolecular decomposition of an interacting pair with an average time τ_c of life of the particles in the cage. According to this assumption the distribution of the duration t_c of the life of the colliding partners in the cage has the form

$$\varphi(t_c)dt_c = \exp(-t_c/\tau_c)dt_c/\tau_c \; . \qquad (2.115)$$

Such a model of the sudden switching on of interaction along with an exponential distribution of the cage lifetimes of particles is also applied in the theory of other bimolecular processes such as triplet-triplet annihilation of excitons [2.63], quenching of triplet states by other paramagnetic additives [2.64], and recombination of radicals in solutions [2.65].

Two particles that encounter each other in a solution and find themselves in a cage undergo a series of recurrent contacts (a detailed discussion of this point is given in [2.26]). The parameter τ_c in (2.115) is, therefore, the total time of all the recurrent contacts of the partners at the exchange radius. Later (see Sect.2.5.4) when discussing the spin exchange between particles with anisotropic interaction, we shall show, by a direct summation of the contributions from all recurrent contacts, that the effective time τ_c of spin exchange (2.115) is indeed contributed to by all the reencounters of the partners in the cage. The contribution of the reencounters is greatly dependent on the kinematics of passing the region of interaction by the particles. The reencounters may be neglected if the motion of the particles is performed by such long jumps that one jump is quite sufficient for the particles to cross the region of interaction. In this case τ_c (2.115) is equal to the average time of one diffusion step of relative movement of two particles. In the other limiting case of the diffusion by sufficiently short jumps the number of reencounters is large. In particular, in the limiting case of continuous diffusion the duration of one contact is infinitely short and only the total time of all the contacts inside a narrow layer of interaction distances has a finite value.

Equations for the Estimation of τ_c

The mean contact time of the colliding particles depends on the parameters of the partners' diffusion and of their interaction. Denote by $D_{A(B)}$, $R_{A(B)}$, $\lambda_{A(B)}$, $\tau_{A(B)}$ the diffusion coefficient, the Van der Waals radius, the mean length of a step, and the mean time of one diffusion step for the molecules A and B, respectively. Assume the hydrodynamic radii of A and B to coincide with their Van der Waals radii. From the theory of Brownian motion it is known that

$$\lambda^2_{A(B)} = 6\, D_{A(B)} \cdot \tau_{A(B)} \quad . \tag{2.116}$$

The mean contact time of particles depends on the kinematics of their passing the region of interaction. For a one-jump passage, i.e., if $\lambda_{A(B)} > R_{A(B)}$ and $\lambda_{A(B)}$ exceed the size of the interaction region, τ_c must be expressed through τ_A and τ_B with the help of the equation

$$1/\tau_c = 1/\tau_A + 1/\tau_B \quad . \tag{2.117}$$

For example, in a collision of two identical paramagnetic particles from (2.116,117) we obtain

$$\tau_c = \lambda_A^2/12\ D_A \quad . \tag{2.118}$$

Assuming the coefficient of diffusion to be linked with a viscosity η of the solvent by the Stokes equation

$$D = kT/6\pi R_h \eta \ , \tag{2.119}$$

where R_h is the hydrodynamical radius of the particle, and taking account of the assumption $R_h = R_A$, from (2.118,119) we obtain

$$\tau_c = \pi R_A \lambda_A^2 \eta/2kT \ , \quad \lambda_A > R_A \quad . \tag{2.120}$$

Consider the opposite situation of the diffusion passage of the interaction region by the particles. In this case it is necessary to take into account all reencounters of the partners. Introduce a narrow layer of interaction located between the spheres with the radii $R_{AB} - a$ and $R_{AB} = R_A + R_B$. From the theory of radical recombination in solutions it is known that in the model of continuous diffusion the total time of the life of the colliding particles in this layer is [2.26,66]

$$\tau_c = aR_{AB}/D_{AB} \ , \quad \text{if} \quad a < R_{AB} \quad . \tag{2.121}$$

In [2.67] it is shown that $a = 1/\varkappa$, if the exchange integral decreases as the distance between the partners increases according to the law $J(r) = J_0 \exp(-\varkappa r)$.

Thus (2.118,121) yield estimates of τ_c in two limiting cases when $\lambda > R_{AB}$ and $\lambda < R_{AB}$. For the intermediate situation $\lambda \sim R_{AB}$ there is, so far, no theory of contacts of particles that would take into account the fact that $a \neq 0$.

Note that in [2.44,61,68] formulas were suggested for estimating τ_c which overestimate its value: e.g., according to [2.61]

$$\tau_c = \pi R_{AB}^3 \eta/kT \quad .$$

In the general case (2.120,121) yield different values of τ_c. For example, if we assume that $\lambda = a = R_A$, then this difference amounts to one order. However, if we accept a relation between the parameters, more reasonable for the jump diffusion model, and assume that $a = R_A/3 = 1$ Å and that a molecule jumps over the intermolecular distance, $\lambda = 2R_A = 6$ Å, then the continuous diffusion and the jump models of passing the exchange range give nearly the same values for τ_c. These considerations show a sufficiently accurate estimate of τ_c to require a detailed knowledge of the mechanism of diffusion. Unfortunately, at present the problem of the mechanism of molecular mobility in a liquid has not yet been finally solved. Further we shall make use of the estimate of τ_c according to (2.121) as from a priori considerations it is easier to estimate the thickness of the interaction layer a than the jump length λ.

The kinematics of diffusion influences not only τ_c, but also the rate constant K_D of diffusion collisions. Below we shall estimate K_D by using the Smolukhovsky equation (2.50) which is valid for the diffusion of uncharged molecules by short jumps or (in the limiting case) for the continuous diffusion in a structureless liquid.

The interaction between partners may essentially change the frequency of diffusion encounters and the contact lifetime. For example, the Coulomb attraction of particles sharply increases τ_c and K_D while the Coulomb repulsive interaction of partners, on the contrary, decreases them. Within the framework of the continuous diffusion model we have the following results [2.26,66] for the unlike-charged particles

$$\begin{aligned} K_D^a &= K_D/w \ , \\ \tau_c^a &= \tau_c w \exp(r^*/R_{AB}) \ , \\ w &= (R_{AB}/r^*)\,[1 - \exp(-r^*/R_{AB})] \ , \\ K_D &= 4\pi R_{AB} D_{AB} \ , \\ \tau_c &= a R_{AB}/D_{AB} \end{aligned} \tag{2.122}$$

and for the like-charged particles

$$\begin{aligned} K_D^r &= K_D^a \cdot \exp(-r^*/R_{AB}) \ , \\ \tau_c^r &= \tau_c^a \cdot \exp(-r^*/R_{AB}) = \tau_c w \ . \end{aligned} \tag{2.123}$$

In these expressions r^* is the Onsager radius, i.e., such a distance at which the Coulomb interaction energy of partners with the charges Z_A and Z_B is equal to the heat energy

$$r^* = |Z_A Z_B| e^2/\varepsilon kT \ .$$

For nonpolar solvents with a low dielectric constant ε the value of r^* may reach hundreds of Ångströms, the Coulomb attraction increasing, possibly by several orders, the time τ_c as compared with its value for uncharged partners.

2.5.2 Kinetic Equations

Within the model of sudden collisions the kinetic equations for describing the processes of spin exchange in diluted solutions were formulated by LYNDEN-BELL [2.69]. Following her work let us divide the spins into two subsystems: isolated paramagnetic particles and dimer pairs of interacting particles in the cage. These two subsystems are in dynamic equilibrium: at a rate $K_D C$ (where C is the spin concentration) a given paramagnetic particle leaves the subsystem of isolated spins and forms a dimer with another paramagnetic molecule, and at a rate $1/\tau_c$ a dimer decomposes into two isolated spins. Before writing kinetic equations one should introduce some notations: $\hat{H}_0(k)$ is the spin-Hamiltonian of an isolated k^{th} paramagnetic molecule; $\hat{V}$ is the spin Hamiltonian of exchange interaction in a dimer; ρ is the isolated spin density matrix; $\rho^{(2)}$ is the dimer density matrix. At the moment when a dimer is formed its density matrix is equal to the external product of the density matrices of the collision partners, i.e.,

$$\rho^{(2)} = \rho \times \rho \quad . \tag{2.124}$$

As particles encounter in a solution, one and the same pair of spins may chance to come into close contact several times. In this case their exchange interaction is successively switched on and off. According to [2.69] the total effect of the exchange interaction in the dynamics of the two spins is assumed to be essentially the same as if the exchange interaction were switched on only once but for the total time of all the contacts. However in fact, in the intervals between the reencounters the state of spins may change due to hfi, paramagnetic relaxation processes, etc. In the general case the dynamics of spins between the contacts may interfere in a rather complicated manner with the exchange interaction. The dynamics of spins in the intervals between the reencounters of the partners has been thoroughly studied in the theory of chemically induced dynamic polarization of nuclei and electrons. This dynamics is shown [2.26] to be capable of changing by 1-10% the probability of radical recombination. The allowance for the spin dynamics in the intervals between reencounters may lead to effects of the same order of magnitude in the spin exchange rate as well.

Within the approximations made the kinetic equations for ρ and $\rho^{(2)}$ take the form [2.69]

$$\begin{aligned} \partial\rho(1)/\partial t &= -i\hbar^{-1}[\hat{H}_0(1), \rho(1)] - K_D C[\rho(1) - \mathrm{Tr}_{1'}\{\rho^{(2)}(1,1')\}]_{1'\to 1} \ , \\ \partial\rho^{(2)}/\partial t &= -i\hbar^{-1}[\hat{H}, \rho^{(2)}] - (\rho^{(2)} - \rho \times \rho)/\tau_c \ ; \quad [\hat{H} = \hat{H}_0(1) + \hat{H}_0(1') + \hat{V}] \ , \end{aligned} \tag{2.125}$$

where Tr is the trace in spin space of one of the collision partners. With the help of these equations an analysis was made of a number of problems of the theory of spin exchange between free radicals [2.70,71] and between triplet excitons [2.56,69].

In the case of spin exchange between unlike particles (2.125) are generalized as follows:

$$\partial\rho^A/\partial t = -i\hbar^{-1}[\hat{H}_A, \rho^A] - K_D C_B(\rho^A - \mathrm{Tr}_B\{\rho^{AB}\}) ,$$
$$\partial\rho^B/\partial t = -i\hbar^{-1}[\hat{H}_B, \rho^B] - K_D C_A(\rho^B - \mathrm{Tr}_A\{\rho^{AB}\}) ,$$
$$\partial\rho^{AB}/\partial t = -i\hbar^{-1}[\hat{H}_A + \hat{H}_B + \hat{V}, \rho^{AB}] - (\rho^{AB} - \rho^A \times \rho^B)/\tau_c \quad . \tag{2.126}$$

In diluted solutions the contact time τ_c is short compared to the time interval between collisions, i.e., $K_D C\ \tau_c \ll 1$, and the impact approximation is applicable. In this situation (2.125) or (2.126) can be considerably simplified [2.23]. Let us solve the equation for the two-particle density matrix ρ^{AB}. We find

$$\begin{aligned}\rho^{AB}(t) &= \exp(-i\hbar^{-1}\hat{H}t)\rho^{AB}(0)\exp(i\hbar^{-1}\hat{H}t)\\ &\times \exp(-t/\tau_c) + \int_0^t \exp(-i\hbar^{-1}\hat{H}\tau)\rho^A(t-\tau)\\ &\times \rho^B(t-\tau)\exp(i\hbar^{-1}\hat{H}\tau)\cdot\exp(-\tau/\tau_c)d\tau/\tau_c ,\end{aligned} \tag{2.127}$$

where $\rho^{AB}(0)$ is the initial density matrix, and $\hat{H} = \hat{H}_A + \hat{H}_B + \hat{V}$. Collisions produce changes in the one-particle matrices ρ with a characteristic time of diffusion between collisions equal to $1/K_D C$. The integrand in (2.127) is different from zero only at $\tau \lesssim \tau_c$. Therefore, to accuracy of small magnitude of about $K_D C\tau_c$ in the integrand one may neglect the changes of ρ due to collisions and, as a result, for short τ one may take

$$\rho^{A(B)}(t-\tau) \simeq \exp(i\hat{H}_{A(B)}\tau/\hbar)\rho^{A(B)}(t)\exp(-i\hat{H}_{A(B)}\tau/\hbar) \quad . \tag{2.128}$$

Substituting this expression into (2.127) and considering that at the kinetic stage of the process, i.e., for $t \sim 1/K_D C \gg \tau_c$, ρ^{AB} must not depend on the initial state, we obtain

$$\rho^{AB}(t) \to \int_0^\infty S(\tau)\rho^A(t) \times \rho^B(t)S^{-1}(\tau) \circ \exp(-\tau/\tau_c)d\tau/\tau_c , \tag{2.129}$$

where we introduce the collision matrix

$$S(\tau) = \exp(-i\hbar^{-1}\hat{H}\tau) \cdot \exp(i\hbar^{-1}\hat{H}_0\tau) \; ;$$
$$\hat{H}_0 = \hat{H}_A + \hat{H}_B \; , \quad \hat{H} = \hat{H}_0 + \hat{V} \quad . \tag{2.130}$$

Substituting (2.129) into (2.126) we obtain the following kinetic equations for ρ^A and ρ^B in the impact approximation

$$\begin{aligned} \partial\rho^A/\partial t &= -i\hbar^{-1}[\hat{H}_A, \rho^A] \\ &\quad -K_D C_B(\rho^A - \mathrm{Tr}_B\{\langle S(\tau)\rho^A \times \rho^B S^{-1}(\tau)\rangle\}) \; , \\ \partial\rho^B/\partial t &= -i\hbar^{-1}[\hat{H}_B, \rho^B] \\ &\quad -K_D C_A(\rho^B - \mathrm{Tr}_A\{\langle S(\tau)\rho^A \times \rho^B S^{-1}(\tau)\rangle\}) \quad . \end{aligned} \tag{2.131}$$

Equations in the form of (2.131) were suggested in [2.72-74].

These equations can be rewritten in a more compact form if we introduce a tensor $\hat{T}$ of collision efficiency equal to $\hat{T} = \langle S(\tau) \times S^{-1}(\tau)\rangle$. The components of $\hat{T}$ are expressed through the elements of the S matrix

$$T_{mn,k\ell} = \int_0^\infty S_{mk}(\tau)S^{-1}_{\ell n}(\tau)\, \exp(-\tau/\tau_c)d\tau/\tau_c \quad . \tag{2.132}$$

Equations (2.131) take the form [2.23]

$$\begin{aligned} \rho^{AB} &= \hat{T}\rho^A \times \rho^B \; , \\ \partial\rho^A/\partial t &= -i\hbar^{-1}[\hat{H}_A, \rho^A] - K_D C_B(\rho^A - \mathrm{Tr}_B\{\hat{T}\rho^A \times \rho^B\}) \; , \\ \partial\rho^B/\partial t &= -i\hbar^{-1}[\hat{H}_B, \rho^B] - K_D C_A(\rho^B - \mathrm{Tr}_A\{\hat{T}\rho^A \times \rho^B\}) \quad . \end{aligned} \tag{2.133}$$

Thus the problem of calculating the efficiency of spin exchange in a single collision (2.132) and that of statistical averaging over collisions in an ensemble of spins (2.133) are separated. This makes it much easier to perform specific calculations and it is this that presents the advantage of kinetic equations (2.131) over (2.126).

The relaxation mechanism of paramagnetic particles isolated in a solution contributes additively to the total relaxation, so one should add the corresponding terms $(\partial\rho^A/\partial t)_{relax}$ and $(\partial\rho^B/\partial t)_{relax}$ to the right-hand sides of (2.131,133). The explicit form of these relaxation terms for various mechanisms has been discussed in detail elsewhere [2.22,28,75].

Besides the exchange interaction there is also the spin-spin dipole-dipole interaction between paramagnetic particles. The generalization of kinetic equations and the calculations of kinetic coefficients for the dipole-dipole mechanism of relaxation are dealt with in detail in [2.34].

The above kinetic equations refer to a situation when the exchange integral is isotropic, i.e., is independent of the mutual orientation of colliding partners. The generalization of these equations for the case of anisotropic exchange interaction was made in [2.23,76]. The generalization of kinetic equations for the systems where one of the collision partners is characterized by comparatively short time of paramagnetic relaxation and where at the moment of the contact of spins it is necessary to take into account the interference between the exchange interaction and the paramagnetic relaxation processes was carried out in [2.23,77,78]. The results of these works will be discussed further. It will also be shown there that from (2.130,131) one can derive the phenomenological kinetic equations that have already been presented in Sect. 2.3.

Steady State Equation for the Operator of Collision Efficiency

To find the components of the tensor $\hat{T}$ one can first find the collision matrix and then with the help of (2.132) calculate the elements of $\hat{T}$. But one can also act differently: find a system of equations directly for the components $T_{mn,k\ell}$ and solve them. Let us find an equation for the operator $\hat{T}$. By differentiating $\hat{T}(\tau) = S(\tau) \times S^{-1}(\tau)$ the operator $\hat{T}(\tau)$ can be shown to satisfy the equation

$$\partial T(\tau)/\partial t = -[\hat{T}(\tau), \hat{Q}] + \hat{\Omega}\hat{T}(\tau) , \tag{2.134}$$

where we introduce the tensors

$$Q_{mn,k\ell} = -i\hbar^{-1}(H_{0_{mk}}\delta_{n\ell} - H_{0_{\ell n}}\delta_{mk}) ,$$

$$\Omega_{mn,k\ell} = -i\hbar^{-1}(V_{mk}\delta_{n\ell} - V_{\ell n}\delta_{mk}) ;$$

$\hat{H}_0 = \hat{H}_A + \hat{H}_B$, δ_{mn} is the Kronecker delta. The efficiency of collision, according to (2.132) is characterized by $\hat{T}(\tau)$ averaged over the duration of a collision. Multiplying (2.134) by $\exp(-\tau/\tau_c)d\tau/\tau_c$, integrating over τ, and taking into account that $\hat{T}(\tau = 0)$ is a unit operator $\hat{E}$ we obtain the desired equation for $\hat{T}$ [2.23]

$$[\hat{T}, \hat{Q}] = \hat{\Omega}\hat{T} - (\hat{T} - \hat{E})/\tau_c . \tag{2.135}$$

Thus, for the components of $\hat{T}$ we obtain a system of algebraic equations with constant coefficients. Equation (2.135) will also be needed for some generalizations of the theory.

2.5.3 Spin Exchange Between Radicals[4]

The objects very convenient for studying spin exchange processes are free radicals. They have relatively long times of paramagnetic relaxation. For solutions of organic radicals T_1, T_2 are usually $\sim 10^{-6}$ s [2.22,34]. Against the background of these sufficiently slow processes one can easily enough reveal and study the spin exchange.

Model System

To begin with, consider the simplest case of spin exchange between two kinds of radicals with different Zeeman frequencies provided the radicals have no magnetic nuclei. We shall dwell in more detail on this problem although in reality free radicals, as a rule, do have magnetic nuclei. Morevoer, it is precisely the collapse of the hfs components of the ESR spectra of radicals that is one of the marked spectral effects of spin exchange and serves as a source of information on the exchange. The suggested model problem, however, is directly related to spin exchange in solutions of free radicals in general. Indeed, the ensemble of radicals can be divided into subensembles belonging to individual hfs components of the ESR spectrum. The entire system looks like a multicomponent system of radicals with different resonance frequencies. The spin exchange in one encounter of two radicals belonging to different hfs components will be described by the results obtained for the model system under consideration.

The spin-Hamiltonians of the Zeeman interaction of spins with the external magnetic field H_0 are

$$\hat{H}_A = g_A \beta H_0 \hat{S}_{Az} \equiv \hbar\omega_A S_{Az} \quad ,$$
$$\hat{H}_B = g_B \beta H_0 \hat{S}_{Bz} \equiv \hbar\omega_B S_{Bz}$$

while the spin-Hamiltonian of exchange interaction is given by (2.1). Then in the presentation of the states $|m_A, m_B\rangle = |m_A\rangle \times |m_B\rangle$ where $|m_A\rangle$ and $|m_B\rangle$ are the eigenvectors of $\hat{S}_{Az}$ and $\hat{S}_{Bz}$ we obtain the following nonzero elements of the S matrix:

$$\langle \tfrac{1}{2},\tfrac{1}{2}|S|\tfrac{1}{2},\tfrac{1}{2}\rangle = \langle -\tfrac{1}{2},-\tfrac{1}{2}|S|-\tfrac{1}{2},-\tfrac{1}{2}\rangle = \exp(-iJ\tau/4) \quad ,$$
$$\langle \tfrac{1}{2},-\tfrac{1}{2}|S|\tfrac{1}{2},-\tfrac{1}{2}\rangle = \exp(i\delta\tau/2) \cdot [b^2 \cdot \exp(-iE_b\tau) + c^2 \cdot \exp(-iE_c\tau)] \quad ,$$

---

4 The discussion that follows holds for any particles with the spin 1/2 and sufficiently long relaxation times.

$$<-\tfrac12,\tfrac12|S|-\tfrac12,\tfrac12> = \exp(-i\delta\tau/2)\cdot[c^2\cdot\exp(-iE_b\tau) + b^2\cdot\exp(-iE_c\tau)] \; ,$$

$$<\tfrac12,-\tfrac12|S|-\tfrac12,\tfrac12> = <-\tfrac12,\tfrac12|S|\tfrac12,-\tfrac12>\cdot\exp(-i\delta\tau)$$

$$= \exp(-i\delta\tau/2)\cdot[\exp(-iE_b\tau) - \exp(-iE_c\tau)]bc \; , \qquad (2.136)$$

where we introduce the notation

$$\delta = \omega_A - \omega_B \; ,$$

$$E_b = -J/4 + R/2 \; , \quad E_c = -J/4 - R/2 \; , \quad R = (J^2 + \delta^2)^{\frac12} \; ,$$

$$\mathrm{tg}\;\varphi = J/\delta \; , \quad b = \cos(\varphi/2) \; , \quad c = \sin(\varphi/2) \; , \quad -\pi/2 \leqq \varphi \leqq \pi/2 \; .$$

With the help of (2.132,136) one can calculate the components of the efficiency operator and thereby find the necessary kinetic coefficients. As a result, from (2.132,133) we obtain

$$\partial\rho^A_{\frac12,\frac12}/\partial t = -K_e C_B(\rho^A_{\frac12,\frac12}\cdot\rho^B_{-\frac12,-\frac12} - \rho^A_{-\frac12,-\frac12}\cdot\rho^B_{\frac12,\frac12}) + 2\,\mathrm{Re}\{K_1\rho^A_{\frac12,-\frac12}\cdot\rho^B_{-\frac12,\frac12}\} \; ,$$

$$\partial\rho^A_{\frac12,-\frac12}/\partial t = -i\omega_A\rho^A_{\frac12,-\frac12} - K_D C_B(1 - p_1\cdot\rho^B_{\frac12,\frac12} - p_2\cdot\rho^B_{-\frac12,-\frac12})\cdot\rho^A_{\frac12,-\frac12}$$

$$+ K_D C_B(p_1'\cdot\rho^A_{\frac12,\frac12} + p_2'\cdot\rho^A_{-\frac12,-\frac12})\cdot\rho^B_{\frac12,-\frac12} \; , \qquad (2.137)$$

where we introduce the notation

$$K_e = 2b^2c^2(1 - <\cos R\tau>)K_D \; ,$$

$$p_1 = <\exp(-iJ\tau/4 + i\delta\tau/2)\cdot[c^2\cdot\exp(iE_b\tau) + b^2\cdot\exp(iE_c\tau)]> \; ,$$

$$p_2 = <\exp(iJ\tau/4 + i\delta\tau/2)\cdot[b^2\cdot\exp(-iE_b\tau) + c^2\cdot\exp(-iE_c\tau)]> \; ,$$

$$p_1' = bc<\exp(-iJ\tau/4 - i\delta\tau/2)\cdot[\exp(iE_b\tau) - \exp(iE_c\tau)]> \; ,$$

$$p_2' = bc<\exp(iJ\tau/4 - i\delta\tau/2)\cdot[\exp(-iE_b\tau) - \exp(-iE_c\tau)]> \; ,$$

$$K_1 = K_D C_B bc<\exp(i\delta\tau)\cdot[b^2 - c^2 - b^2\cdot\exp(-iR\tau) + c^2\cdot\exp(iR\tau)]> \; . \qquad (2.138)$$

Similar equations are also valid for ρ^B.

For real experimental conditions the equations obtained permit essential simplifications. The spin exchange in solutions is studied under the conditions when the system is close to equilibrium. Let us introduce matrices σ which describe a deviation of a system of spins from equilibrium

$$\rho^A = \rho^A_0 + \sigma^A \; , \quad \rho^B = \rho^B_0 + \sigma^B \; . \qquad (2.139)$$

Substituting (2.139) into (2.137) and retaining only the terms linear in σ one can obtain linearized equations which describe the kinetics of spin exchange under the conditions close to an equilibrium. Besides, for typical experimental conditions $\delta \cdot \tau_c \ll 1$, and in the process of a collision one can neglect the fact that spins are nonresonant. Linearizing (2.137) and assuming in (2.138) $\delta \cdot \tau_c \ll 1$, we come to the equations

$$\begin{aligned}(\partial/\partial t)\sigma^A_{\frac{1}{2},\frac{1}{2}} &= -K_e C_B(\sigma^A_{\frac{1}{2},\frac{1}{2}} \cdot \rho^B_{0_{-\frac{1}{2},-\frac{1}{2}}} - \sigma^A_{-\frac{1}{2},-\frac{1}{2}} \cdot \rho^B_{0_{\frac{1}{2},\frac{1}{2}}} \\ &\quad - \sigma^B_{\frac{1}{2},\frac{1}{2}} \cdot \rho^A_{0_{-\frac{1}{2},-\frac{1}{2}}} + \sigma^B_{-\frac{1}{2},-\frac{1}{2}} \cdot \rho^A_{0_{\frac{1}{2},\frac{1}{2}}}) \ , \\ (\partial/\partial t)\sigma^A_{\frac{1}{2},-\frac{1}{2}} &= i[-\omega_A - K_D C_B J\tau_c(\rho^B_{0_{\frac{1}{2},\frac{1}{2}}} - \rho^B_{0_{-\frac{1}{2},-\frac{1}{2}}})/2(1 + J^2\tau_c^{\,2}) \\ &\quad + (\tfrac{1}{2})\delta\tau_c K_e C_B]\sigma^A_{\frac{1}{2},-\frac{1}{2}} - K_e C_B(\sigma^A_{\frac{1}{2},-\frac{1}{2}} - \sigma^B_{\frac{1}{2},-\frac{1}{2}}) \ .\end{aligned} \tag{2.140}$$

From these equations the spin exchange is seen to cause a shift of the resonance frequency of spins and their phase relaxation. The resonance fequency shift is due to two reasons. The shift equal to

$$K_D C_B[J\tau_c/2(1 + J^2\tau_c^{\,2})] \cdot (\rho^B_{0_{\frac{1}{2},\frac{1}{2}}} - \rho^B_{0_{-\frac{1}{2},-\frac{1}{2}}}) \tag{2.141}$$

is proportional to the difference of the state populations of the partner-spin, i.e., to the polarization of the latter. This shift results from the fact that the spins, due to their prevalent orientation to one side, create on the site of the interacting partners a certain average nonzero field of exchange interaction. Under the ordinary conditions, however, when $H_0 \approx 3000$ G, $T \approx 300$ K, the equilibrium polarization of spins is small, $\rho_{0\frac{1}{2},\frac{1}{2}} - \rho_{0-\frac{1}{2},-\frac{1}{2}} \sim 10^{-3}$. This shift is therefore negligible in experiments where no spin polarization of the partners is expressly carried out. The shift in question was studied experimentally for electrons in gases participating in spin exchange with rubidium atoms [2.79]. The spins of Rb atoms were polarized by optical pumping. The ratio of the shift to the value of the exchange broadening amounted to 0.025. Similar measurements in the system electrons + Cs atoms were made [2.80]. In the case of the optical pumping of the collision partner its polarization is essentially larger than the equilibrium value, so the frequency shift in question can be observed experimentally.

In addition, according to (2.140) a frequency shift equal to $(1/2)\delta\tau_c K_e C_B$ occurs. It may be interpreted as being due to the fact that during a collision the unpaired electrons of both partners are equally distributed between the orbitals that have frequencies ω_A and ω_B. Indeed, $\tau_c K_e C_B$ is a fraction of the time spent by the molecule A in contact with B. On the average, therefore, the resonance frequency of the spin of A is equal not to ω_A, but to

$$\begin{aligned}\bar{\omega}_A &= \omega_A(1 - \tau_c K_e C_B) + \tau_c K_e C_B(\omega_A + \omega_B)/2 \\ &= \omega_A - \delta\tau_c K_e C_B/2 \quad . \end{aligned} \tag{2.142}$$

For free radicals $\delta\tau_c \ll 1$, so the line shift $(1/2)\delta\tau_c K_e C_B$ appears to play a negligible part as well. However, in some specific systems the line shift may be observed experimentally. It may be essential, e.g., for spin exchange between triplet excitons or in the case of exchange between radicals and complexes of transition metals for which the parameter δ can be far larger than in the case of exchange between radicals or for the unlike-charged partners when τ_c is large enough. It is of interest to note that the frequency shift $\delta\tau_c K_e C_B/2$ moves the absorption line in the ESR spectrum towards its center, and grows linearly with spin concentration. Phenomenologically, this line shift manifests itself just as the one [see (2.88)] due to the interference of the equivalent spin exchange with other mechanisms of transverse spin relaxation. It should be borne in mind, therefore, that in the general case the line shift in the slow exchange situation which is linear in spin concentration may be due to several causes [see (2.88,142)].

Assuming in (2.140) $\rho_{0_{nn}} = 1/2$ and $\delta\tau_c \ll 1$ we obtain the equations which have found wide application in interpeting the data on spin exchange between radicals [2.68,71]

$$\begin{aligned}(\partial/\partial t)\sigma^A_{\frac{1}{2},\frac{1}{2}} &= -(\tfrac{1}{2})K_e C_B(\sigma^A_{\frac{1}{2},\frac{1}{2}} - \sigma^A_{-\frac{1}{2},-\frac{1}{2}} - \sigma^B_{\frac{1}{2},\frac{1}{2}} + \sigma^B_{-\frac{1}{2},-\frac{1}{2}}) \quad , \\ (\partial/\partial t)\sigma^A_{\frac{1}{2},-\frac{1}{2}} &= -i\omega_A\sigma^A_{\frac{1}{2},-\frac{1}{2}} - K_e C_B(\sigma^A_{\frac{1}{2},-\frac{1}{2}} - \sigma^B_{\frac{1}{2},-\frac{1}{2}}) \quad . \end{aligned} \tag{2.143}$$

Introducing the macroscopic magnetizations of spins via (2.29)

$$\begin{aligned}M_z &= \hbar\gamma(\rho_{\frac{1}{2},\frac{1}{2}} - \rho_{-\frac{1}{2},-\frac{1}{2}})C/2 \quad , \\ M_- &= \hbar\gamma\rho_{\frac{1}{2},-\frac{1}{2}} \cdot C \quad , \end{aligned}$$

we derive from (2.143) the equations introduced in Sect. 2.3 [see (2.56,60)]. In the case under consideration, when $\delta\tau_c \ll 1$, there occurs an equivalent spin exchange with the rate constant

$$\begin{aligned}K_e &= pK_D \quad , \\ p &= (\tfrac{1}{2})J^2\tau_c^2/(1 + J^2\tau_c^2) \quad . \end{aligned} \tag{2.144}$$

This result was first obtained by CURRIN [2.68] and was later reproduced by many authors.

Depending on the scale of exchange interaction the collisions may be classified into weak and strong. Collisions are referred to as weak if

$$J\tau_c < 1 \quad . \tag{2.145}$$

In this case the exchange rate constant is

$$K_e \cong (\tfrac{1}{2})J^2\tau_c{}^2K_D \sim J^2/D \quad . \tag{2.146}$$

It displays a strong dependence on the value of the exchange integral and is inversely proportional to the coefficient of mutual diffusion of colliding particles.

For strong collisions when

$$J\tau_c > 1 \tag{2.147}$$

the exchange rate no longer depends on J and is determined only by the frequency of diffusion collisions

$$K_e \cong K_D/2 \sim D \quad . \tag{2.148}$$

The kinetics of weak exchange is controlled by the efficiency of collisions while that of strong exchange, by diffusion encounters. Note that these two extreme situations characterize any bimolecular processes.

Thus, when the exchange integral between adjacent paramagnetic molecules is large compared to the difference of their resonance frequencies, i.e., $J \gg \delta$ and $\delta\tau_c \ll 1$, then according to (2.143) an equivalent spin exchange occurs. In reality, as a rule, it is only this case that is of practical interest. Therefore we shall not dwell much on the opposite situation when $J < \delta$. In this case the exchange interaction cannot compensate the change of the Zeeman energy of spins as a result of their flip-flop, so the latter occurs only with a small probability proportional to $J^2\tau_c^2/(1+\delta^2\tau_c^2)$. The principal mechanism of phase relaxation is now spin dephasing rather than spectral diffusion induced by equivalent spin exchange. Indeed, at $J < \delta$ from (2.137,138) we obtain an equation in which the term describing the spin state exchange with a partner turns out to be negligible [cf with (2.43)]

$$(\partial/\partial t)\sigma^A_{\frac{1}{2},-\frac{1}{2}} = -i\omega_A\sigma^A_{\frac{1}{2},-\frac{1}{2}} - [J^2\tau_c^2/4(1 + J^2\tau_c^2/4)] \cdot K_D C_B \sigma^A_{\frac{1}{2},-\frac{1}{2}} \quad . \tag{2.149}$$

For organic free radicals the difference of the g values is of the order of 10^{-2} - 10^{-3}, so the typical values of $\delta \sim 10^8$ rad/s and the difference of the Zeeman frequencies of radicals does not practically play any role in the exchange process. However, such differences could be of non-negligible significance in the exchange between the radicals and paramagnetic ions. Indeed, the components of the g tensor of paramagnetic complexes may be very different from those of the g tensor of the free electron, and in fields of $H_0 \approx 3000$ G sufficiently large values of $\delta \sim 10^{11}$ rad/s may be realized.

Comparison of (2.143) with (2.149) shows that the latter contains no term expressing the response of transverse spin polarization from the collision partner. Only a spin dephasing occurs. The mechanism of spin dephasing at $J < \delta$ is as follows. At the moment of collision the exchange field of the collision partner is added to the external field and the spin of A precesses at the frequency $\omega_A + Jm_B$ (m_B is the spin projection of the collision partner) rather than at ω_A. As a result the collision changes the spin phase by the value of $Jm_B t_c$.

Spin Exchange Between Radicals in the Presence of Hyperfine Interaction

The results obtained for the spin exchange between two kinds of radicals are directly generalized for the case of exchange in the solutions of radicals which are characterized by a hfs of ESR spectra. For example (2.133) turns into [2.23]

$$\partial\rho_k/\partial t = -i\hbar^{-1}[\hat{H}_k, \rho_k] - K_D C \sum_{n(\neq k)} \varphi(n)(\rho_k - \mathrm{Tr}_n\{\hat{T}\rho_k \times \rho_n\}) , \qquad (2.150)$$

where ρ_k, $\hat{H}_k$, and $\varphi(k)$ are the partial density matrix, the spin-Hamiltonian, and the statistical weight of the radicals belonging to the kth hfs component. By repeating the reasoning followed in obtaining (2.143) and introducing the macroscopic magnetic moments of the components from (2.150) we come to the phenomenological equations (2.65). The exchange rate constant K_e is given by (2.144) for any hfs of the ESR spectrum.

2.5.4 Spin Exchange Between Radicals with Anisotropic Interaction

The exchange rate constant (2.144) refers to the case of isotropic interaction when J is independent of the mutual orientation of colliding partners. However the exchange interaction is often anisotropic. This is connected with the anisotropy of the orbitals occupied by unpaired electrons. According to (2.9) the exchange integral is proportional to the square of the overlap integral. Then, e.g., for a pair of electrons, one of which occupies an s orbital and the other a p orbital, $J = J_\sigma \cdot \cos^2\theta$ (here J_σ is the exchange integral for the σ bond of the orbitals and θ is the angle between the axis of the p orbital and the line between the colliding partners).

The anisotropy of spin exchange can be averaged by the relaxation of the mutual orientation of the partners during their contact. In a solution one and the same pair of particles undergoes a series of reencounters. Hence there is another possibility of averaging the anisotropy of spin exchange due to the orientational relaxation of the partners in the intervals between their reencounters.

Rotation of Partners at the Moment of Their Contact

Within the framework of the sudden collisions model let us first consider the averaging of the anisotropy resulting from the rotation of the partners at their contact.

The influence of the anisotropy of interaction on spin exchange depends on the relation between the contact lifetime τ_c and the time τ_0 of orientational relaxation. If $\tau_0 >> \tau_c$, then in each collision the exchange integral has quite a definite fixed value. The spin exchange efficiency summarized over all possible collisions is equal to

$$p = (1/2) \int dJ \cdot w(J)J^2\tau_c{}^2/(1 + J^2\tau_c{}^2) \; , \qquad (2.151)$$

where $w(J)dJ$ is the statistical weight of collisions for which J fits into the interval $(J, J + dJ)$.

For radicals whose spin density is localized, say, only on the peripheral atom, the distribution $w(J)$ can be chosen in the form

$$w(J) = f\delta(J - J_0) + (1 - f)\delta(J) \; , \qquad (2.152)$$

where $\delta(\;)$ is a delta function. Thus there is a probability f that at contact $J \neq 0$, while in the other collisions the exchange interaction is zero. Then we have

$$p = (1/2)\, fJ_0^2\tau_c^2/(1 + J_0^2\tau_c^2) \qquad (2.153)$$

so that the rate constant of spin exchange, as distinct from (2.144), contains an additional steric factor

$$K_e = f \cdot (1/2) \cdot [J_0^2\tau_c^2/(1 + J_0^2\tau_c^2)]K_D \; . \qquad (2.154)$$

Thus, in this case, when $\tau_0 >> \tau_c$ and the distribution w(J) is given by (2.152), the anisotropy effect of interaction can be described by introducing a steric factor.

In the general case even at $\tau_0 >> \tau_c$ the anisotropy effect of interaction cannot completely be accounted for by introducing a steric factor. In particular there may be an essential change in the temperature dependence of the exchange rate. Indeed, when there is a set of values of J, some collisions may lead to a weak spin exchange, the others to a strong one. Thus simultaneously collisions whose contributions to K_e vary inversely with temperature [cf (2.146,148)] may occur.

The existence of the set of J in a collision can be established by analyzing the dependence of the exchange rate constant on viscosity or temperature. Indeed, according to (2.151) the width of the intermediate region between weak exchange when $J\tau_c < 1$ and strong exchange when $J\tau_c > 1$ depends directly on the width of the distribution function w(J).

In the opposite case of very fast rotations when $\tau_0 << \tau_c$ it is evident that only the mean value of the exchange integral $\langle J\rangle = \int dJ \cdot Jw(J)$ is reflected in the exchange. The exchange rate constant is given by the expression

$$K_e = (1/2)[\langle J\rangle^2\tau_c^2/(1 + \langle J\rangle^2\tau_c^2)]K_D \tag{2.155}$$

which coincides with (2.144) if J is replaced by its mean value.

The exchange interaction of a pair of colliding paramagnetic particles can also be modulated by their oscillations. Taking into account that the period of intramolecular oscillations is short compared to the time τ_c one can affirm that in this case K_e will be described by (2.155) as well.

Rather complex both for theoretical analysis and for comparing theory with experiment is the intermediate case $\tau_0 \sim \tau_c$. In this case the partners succeed in changing their mutual orientation within the time of collision. When calculating the exchange efficiency one must therefore take into account the random modulation of the exchange integral. The effect of particle rotations on the spin exchange rate was first discussed by EDELSTEIN et al. [2.81]. Regarding spin exchange between partners in a cage as a unimolecular process, the authors of [2.81] suggested for exchange efficiency the formula $p = 1 - \exp(-J\tau_c)$ where, according to [2.81], J is the mean value of the exchange integral. In this form, however, their formula is wrong. It would be more correct, by analogy with unimolecular chemical reactions, to represent p in the form $p = 1 - \exp(-k_e' t_c)$ where k_e' is the exchange rate constant at the moment of contact. However, k_e' must depend on the exchange interaction in a more complicated manner than is assumed in [2.81]. By averaging over t_c for p we arrive at the result

$$p = k_e'\tau_c/(1 + k_e'\tau_c) \quad .$$

If the limiting value of p in the case of strong exchange does not tend to 1, but to $p_{max} < 1$, then

$$p = p_{max} \cdot k_e'\tau_c/(1 + k_e'\tau_c) \quad . \qquad (2.156)$$

In the case of weak exchange when $k_e'\tau_c < 1$

$$p = p_{max} \cdot k_e'\tau_c \ , \qquad (2.157)$$

i.e., p grows with τ_c linearly rather than quadratically as was the case in (2.146).

The approach reflected in the above two formulae is essentially different from all previous considerations. In the previous discussions the elementary act of spin exchange between two partners that chanced to come into contact was assumed to be due to a dynamic process, i.e., to oscillations between initial and final states, when the changes in the spin states of partners are not described by the usual kinetic model. As for the last two formulae, however, they refer to the situation when the collision time is sufficient for the exchange rate constant to be formed, i.e., within the times τ_c the exchange between two partners enters the kinetic stage with the rate constant k_e'. According to the current concept of the theory of irreversible processes [2.82] the spin exchange between two colliding partners can be described by a kinetic model if there is a fast enough process forming the kinetics of the exchange within the contact time. This process can be the fast rotation of molecules if the exchange interaction is anisotropic, the fast spin relaxation of one of the colliding partners (see Sect. 2.5.5).

A consistent theory of the effect of rotational diffusion of particles in the process of spin exchange was developed in [2.23,76]. It is appropriate to discuss these studies in some detail. They are not only useful in analyzing the effects connected with the rotation of particles but will also be useful in discussing other mechanisms of the random modulation of exchange interaction of particles (see Sects.2.5.5 and 2.6).

If J varies randomly so that the exchange interaction is described as

$$\hat{V}(t) = \hbar J(t)\hat{S}_A\hat{S}_B \qquad (2.158)$$

then the equation for the efficiency tensor (2.134) takes the form

$$\partial\hat{T}(\tau)/\partial\tau = -[\hat{T}(\tau), \hat{Q}] + \hat{\Omega}(\tau)\hat{T}(\tau) \tag{2.159}$$

in which, contrary to (2.134), the interaction tensor components $\hat{\Omega}(\tau)$ are randomly modulated by the rotation of particles

$$\Omega_{mn,k\ell}(\tau) = -i\hbar^{-1}[\hat{V}_{mk}(\tau)\delta_{n\ell} - \hat{V}_{\ell n}(\tau)\delta_{mk}] \quad . \tag{2.160}$$

In order to find the desired tensor $\hat{T}$ of the collision efficiency which is introduced into the kinetic equations (2.150) one should solve (2.159) whose coefficients vary randomly in time, and then perform two averagings 1) over all possibilities of the random process J(t) and 2) over all possible duration of collision. The random variations of particle orientation are described fairly well by a Markovian process. Hence the parameter J(t) in (2.158) can also be considered as a Markovian process. This circumstance permits the averaging of (2.159) over all possible values of J(t). The procedure of averaging equations of the type of (2.159) is described, e.g., in [2.23,32,33]. Introduce a partial tensor $\hat{T}(\tau,J)$ of a subensemble of dimers (pairs of contacting particles) for which at a moment τ an assigned value of the exchange integral J occurs. The tensor $\hat{T}(\tau,J)$ varies for two reasons. On the one hand, the spin state changes under the influence of their interactions. This variation of $\hat{T}(\tau,J)$ is described by (2.134) with a corresponding value of J. On the other hand, in the course of rotational diffusion some dimers leave the subensemble with a chosen value of J, while some others enter it. The total variation of $\hat{T}(\tau,J)$ is described by the equation

$$(\partial/\partial\tau)T(\tau,J) = -[T(\tau,J), \hat{Q}] + \hat{\Omega}(J)\hat{T}(\tau,J) + [(\partial/\partial\tau)\hat{T}(\tau,J)]_0 \quad . \tag{2.161}$$

A specific form of the last term here depends on how the rotational diffusion occurs and how the exchange integral is connected with the mutual orientation of colliding partners. Multiplying (2.161) by $\exp(-\tau/\tau_c)d\tau/\tau_c$ and integrating over τ, one can find a stationary equation for the partial tensor $\hat{T}(J)$ of collision efficiency. The desired tensor $\hat{T}$ of efficiency will be given by the relation

$$\hat{T} = \int dJ \cdot w(J)\hat{T}(J) \quad . \tag{2.162}$$

As pointed out above, for free radicals at the moment of collision one can neglect effects connected with the difference of their g values or with hfi. Then the Hamiltonians $\hat{H}_0$ and $\hat{V}$ commute and have common eigenvectors. In the representation of the eigenfunctions of $\hat{V}$ from (2.159) we obtain the following equations for the components of $\hat{T}(\tau)$

$$(\partial/\partial\tau)T_{mn,k\ell}(\tau) = -i\hbar^{-1}[V_{mm}(\tau) - V_{nn}(\tau)]T_{mn,k\ell}(\tau) \quad . \tag{2.163}$$

These equations give nonzero solutions only for diagonal elements $T_{mn,mn}$ as at $\tau = 0$ $\hat{T}(\tau)$ is a unit tensor.

Solving (2.163) we come to the result

$$T_{mn,mn} = \exp\{-i\hbar^{-1} \int_0^\tau [V_{mm}(t) - V_{nn}(t)]dt\} \quad .$$

Substituting this efficiency tensor $\hat{T}$ into (2.150) and linearizing the resulting equations in the high-temperature approximation $\rho_{0nn} \approx 1/2$ we obtain the following expression for the spin exchange efficiency:

$$p = (1/2)\ \mathrm{Re}\{1 - \int_0^\infty \langle\exp[i \int_0^\tau J(t)dt]\rangle \exp(-\tau/\tau_c)d\tau/\tau_c\} \quad . \tag{2.164}$$

The averaging of a quantity of the type of $\exp[i \int_0^\tau J(t)dt]$ over all realizations of the random process $J(t)$ has been already met with in discussing the spectral manifestations of spin exchange [see (2.92)]. Making use of the same considerations that led to (2.94), in the limiting case of fast rotations, i.e., under the applicability of the stochastic perturbation theory [2.31] with $\tau_0 < \tau_c$ and $d\tau_0\tau_c < 1$ where $d = \langle(J - \langle J\rangle)^2\rangle$, we obtain

$$p = (1/2)\,[\langle J\rangle^2\tau_c^2 + d\tau_0\tau_c(1 + d\tau_0\tau_c)] \times [\langle J\rangle^2\tau_c^2 + (1 + d\tau_0\tau_c)^2]^{-1} \quad . \tag{2.165}$$

Hence the above result (2.155) follows, if $\langle J\rangle^2\tau_c^2 > d\tau_0\tau_c$. If, as a result of random modulation, the exchange interaction is averaged to zero, then the exchange efficiency is determined by the mean-square deviation

$$p = (1/2)\langle J^2\rangle\tau_0\tau_c/(1 + \langle J^2\rangle\tau_0\tau_c) \quad . \tag{2.166}$$

It is unlikely that due to the rotation of partners the exchange integral is averaged to zero. Nevertheless, (2.166) is of certain interest as it holds for any mechanism of the random modulation of exchange interaction. In particular, random reorientations of the spins of partners average the exchange interaction to zero. We shall meet such a situation when we discuss the exchange between radicals and other species with short times of paramagnetic relaxation.

In a situation obeying (2.166) during contact of partners one might introduce an exchange rate constant $k_e' = \langle J^2\rangle\tau_0$ [cf (2.156) with (2.166)].

Accordingly, in the case of weak exchange from (2.166) we obtain

$$p \cong (1/2)k_e'\tau_c = (\tfrac{1}{2})\langle J^2\rangle\tau_0\tau_c \quad .$$

Thus, the exchange rate constant no longer depends on the coefficient of translational diffusion and proves to be inversely proportional to the coefficient of rotational diffusion D_0

$$K_e \cong (1/2)\langle J^2\rangle\tau_0\tau_c K_D \sim \langle J^2\rangle/D_0 \quad . \tag{2.167}$$

To establish the influence of rotation on the efficiency of exchange in the intermediate cases $\tau_0 \sim \tau_c$, it is necessary to specify the random variations of J(t). By way of example let us consider a model of molecule rotation by jumps to relatively large angles. Assume that at a mean frequency $1/\tau_0$ a sudden switching of J occurs while the probability of changing the state with J_1 into that with J_2 is independent of the initial state and equal to the statistical weight of the final state, i.e., $w(J_2)$. Denote by $X(\tau,J)$ the mean value of $\exp[-i\int_0^\tau J(t)dt]$ for such a selection of realizations of the random process J(t) for which J at a moment τ assumes an assigned value. It should be emphasized that here account is taken only of the influence of the rotation of molecules on the evolution of spins while the principally possible inverse effect, dependence of the rotation on the spin state of partners, is neglected. For the chosen model of random changes of J from (2.160) it follows that [cf with (2.65,66)]

$$(\partial/\partial\tau)X(\tau,J) = -iJX(\tau,J) - [X(\tau,J) - \int X(\tau,J)w(J)dJ]/\tau_0 \quad . \tag{2.168}$$

The efficiency of collisions is

$$p = (1/2)\ \mathrm{Re}\{1-\int_0^\infty d\tau \int dJ \cdot X(\tau,J)w(J)\exp(-\tau/\tau_c)/\tau_c\} \quad . \tag{2.169}$$

The solution of the system of equations of the type of (2.168) was obtained in Sect.2.4 [see (2.74-77)]. Thus we arrive at

$$p = (1/2)\ \mathrm{Im}\{i + iG\tau_0/(\tau_0\tau_c + G\tau_c)\} \quad , \tag{2.170}$$

where we introduce the notation

$$G = -\int dJ \cdot w(J)\tau^*/(1 + iJ\tau^*) \quad , \quad \tau^* = \tau_0\tau_c/(\tau_0 + \tau_c) \quad .$$

Especially simple results are obtained in the case when the distribution of J is given by (2.152). For this situation from (2.170) we obtain the following spin exchange efficiency:

$$p = (1/2) f_e J_e^2 \tau_c^2/(1 + J_e^2 \tau_c^2) \ , \qquad (2.171)$$

where

$$f_e = f\chi \ , \quad J_e = J/\chi \ , \quad \chi = (\tau_o + \tau_c)/(\tau_o + f\tau_c) \leqq 1 \ .$$

Thus, with due regard to the orientational relaxation of partners the anisotropic spin exchange may be specified by an effective steric factor which is χ times larger than the geometric one f, and by an effective exchange integral which is χ times less than the integral J under a favorable mutual orientation of partners.

How essential the averaging of the anisotropy of spin exchange will prove to be depends on the "number" of rotations of partners within the time τ_c, i.e., on the value of $n_o = \tau_c/\tau_o$. For partners with pronounced anisotropy of interaction when $f \lesssim 0.1$

$$\chi \cong 1 + n_o \qquad (2.172)$$

the effective steric factor increases proportionally to n_o.

The value of n_o depends on a specific model of rotational and translational diffusion of particles, and on the strength of the interaction of particles. In the simplest case when the interaction between the particles is small compared to kT and the translational diffusion is considered in the framework of the model of continuous diffusion we obtain $n_o \lesssim 1$. Indeed, in this situation $\tau_c = aR_{AB}/D_{AB}$ (2.121). The time of the orientational relaxation of partners may be estimated by the formula

$$1/\tau_o = 1/\tau_{oA} + 1/\tau_{oB} \ , \qquad (2.173)$$

where τ_{oA} and τ_{oB} are the times of rotational relaxation of the partners in a pair. These can be calculated by the Debye formula for the orientational relaxation of molecules in a liquid. Assuming, for simplicity, the radii of the partners to be equal we obtain

$$\tau_o = R_{AB}^2/6D_{AB} \ . \qquad (2.174)$$

As a result, the number of rotations of the partners within their contact lifetime turns out to be equal to

$$n_o = 6a/R_{AB} \ . \qquad (2.175)$$

For typical values of the parameters $a \simeq 0.2-0.3$ Å and $R_{A,B} \simeq 2-5$ Å we obtain $n_0 \lesssim 1$. Thus, the orientational relaxation of contacting partners increases f_e by no more than twofold if the forces of attraction between the particles do not extend their contact time [see, e.g., (2.122)].

Orientational Relaxation of Particles in Intervals Between Reencounters [2.83]

In order to consistently include into consideration the orientational relaxation of particles in the intervals between the reencounters one can act in two ways. One can generalize the kinetic equations for the spin density matrix [2.26]. Denote by $\rho^{AB}(\underline{r},\theta,t)$ the density matrix of the subensemble of pairs with an assigned radius vector $\underline{r}$ between the particles and with their assigned mutual orientation which is determined by the angles $\{\theta\}$. The kinetic equation for $\rho^{AB}(\underline{r},\theta,t)$ may be written in the form [2.26]

$$\partial\rho^{AB}/\partial t = (\hat{Q}_A + \hat{Q}_B)\rho^{AB} - i\hbar^{-1}[\hat{V}(\underline{r},\theta), \rho^{AB}] + (\hat{L}_t + \hat{L}_o)\rho^{AB} , \qquad (2.176)$$

where the first term describes the variation of ρ^{AB} due to the spin motion of the isolated partners A and B, $\hat{V}(\underline{r},\theta)$ is the Hamiltonian of anisotropic exchange interaction, and the last terms describe the variation of $\rho^{AB}(\underline{r},\theta,t)$ due to the translational and rotational diffusion of partners. This equation should be solved unter the following boundary conditions. The condition for large values of r is provided by the assumption of the weakening of interparticles correlation [2.82]

$$\rho^{AB} \xrightarrow[r\to\infty]{} \rho^A \times \rho^B \quad .$$

The condition for small values of r is obtained by requiring the zero value for the flux of ρ^{AB} through the sphere with the radius of the closest approach of the particles [2.84].

Considerations similar to those given in Sect.2.5.2 show that at the kinetic stage of the process we have

$$\begin{aligned}
&\rho^{AB}(\underline{r},\theta,t) \Rightarrow \hat{T}(\underline{r},\theta)\rho^A(t) \times \rho^B(t) , \quad [\hat{T},\hat{Q}_A + \hat{Q}_B] = \hat{\Omega}\hat{T} + \hat{L}_t\hat{T} + \hat{L}_o\hat{T} ,\\
&\partial\rho^A/\partial t = \hat{Q}_A\rho^A + C_B \mathrm{Tr}_B\{ \int d^3\underline{r}\, d\theta\, \hat{\Omega}(\underline{r},\theta)T(\underline{r},\theta)\rho^A \times \rho^B\} ,\\
&\partial\rho^B/\partial t = \hat{Q}_B\rho^B + C_A \mathrm{Tr}_A\{ \int d^3\underline{r}\, d\theta\, \hat{\Omega}(\underline{r},\theta)\hat{T}(\underline{r},\theta)\rho^A \times \rho^B\} .
\end{aligned} \qquad (2.177)$$

Solving these equations for the given model of particle diffusion one can find the rate constant of anisotropic spin exchange.

In the framework of the sudden collisions model one can act differently, i.e., directly sum up the contributions of all reencounters. The method of summing up these contributions was developed for the recombination of free radicals [2.85,86] and generalized to spin exchange [2.83]. Denote by $\hat{F}(t)$ the spin evolution operator of a pair of particles in the interval between contacts, and by $\hat{Q}(t)$ the spin evolution operator of a contacting pair when spin interaction is switched on. The distribution of pairs over the values of their contact lifetimes and the times between their contacts is given by the functions w(t) and f(t), respectively. The density matrix of the subensemble of pairs that had gone through n reencounters and then escaped from a cage is equal to

$$\rho_n^{AB} = (1 - \bar{p})Q(FQ)^n \rho^A \times \rho^B \quad , \tag{2.178}$$

where $\bar{p}$ is the probability of a reencounter of particles, and

$$Q = \int_0^\infty \hat{Q}(t)w(t)dt \ ,$$
$$F = \int_0^\infty \hat{F}(t)f(t)dt \quad . \tag{2.179}$$

Taking account of all reencounters we obtain

$$\rho^{AB} = \sum_n \rho_n^{AB} = (1 - \bar{p})Q(1 - FQ)^{-1} \rho^A \times \rho^B \quad . \tag{2.180}$$

The change in the density matrix of the pair as a result of all reencounters is

$$\Delta\rho^{AB} = [1-(1 - \bar{p})Q(1 - FQ)^{-1}]\rho^A \times \rho^B \equiv \hat{T}\rho^A \times \rho^B \quad . \tag{2.181}$$

The change in the density matrix of each of the partners as a result of their encounter is

$$\Delta\rho^A = Tr_B\{\hat{T}\rho^A \times \rho^B\} \ ,$$
$$\Delta\rho^B = Tr_A\{\hat{T}\rho^A \times \rho^B\} \quad . \tag{2.182}$$

For the conditions of experimental study of spin exchange (2.182) may be linearized with the help of (2.139) and, as a result, one can find the efficiency of spin exchange per encounter in a solution.

Such a summation of contributions of all reencounters was carried out in [2.83] for a relatively simple model: the anisotropy of exchange interaction is characterized by a steric factor f which is equal to the statistical weight of mutual particle orientations favorable for spin exchange. In the intervals between contacts of particles a relaxation of their mutual orientation occurs.

The kinetics of this process is described in [2.83] by adopting the simplest model

$$\partial\rho_1^{AB}/\partial t = -(1 - f)\rho_1^{AB}/\tau_0' + f\rho_2^{AB}/\tau_0' ,$$
$$\partial\rho_2^{AB}/\partial t = (1 - f)\rho_1^{AB}/\tau_0' - f\rho_2^{AB}/\tau_0' . \tag{2.183}$$

Here ρ_1^{AB} and ρ_2^{AB} are the density matrices of the ensembles of pairs with the mutual orientation of partners favorable and unfavorable for spin exchange, respectively. The orientational relaxation occurs as a result of rotational and translational diffusion of the partners in a pair. The rotation of particles seems to make the main contribution, and τ_0' in (2.183) may be estimated by means of (2.173,174). The variations of ρ^{AB} in the intervals between contacts can be expressed through the linear operator

$$\rho^{AB}(t) = \hat{F}(t)\rho^{AB}(0) . \tag{2.184}$$

An explicit form of $\hat{F}$ is found from a solution of (2.183). The operator $\hat{F}(t)$ is averaged according to (2.179) with the help of the function of distribution of reencounters which was obtained by NOYES [2.87]

$$f(t) = mt^{-3/2} \cdot \exp(-\pi m^2/\bar{p}^2 t) ,$$
$$m = (27/8\pi)^{\frac{1}{2}} \cdot (1 - \bar{p})^2 \cdot (R_{AB}/\lambda)^2 \cdot \tau^{\frac{1}{2}} , \tag{2.185}$$

where λ and τ are the length and time of an elementary act of the relative displacement of the partners. The probability of a reencounter depends on the relation between λ and $R_{AB} = R_A + R_B$. At $\lambda \sim R_A + R_B$ $\bar{p} \cong R_{AB}/(R_{AB} + \lambda)$ [2.26], and at $\lambda \ll R_{AB}$ [2.87]

$$\bar{p} \cong 3R_{AB}/(3R_{AB} + 2\lambda) . \tag{2.186}$$

At the moment of contact of partners, besides the orientational relaxation, a spin exchange occurs. Therefore at the moment of contact the kinetic equations take the form

$$\partial\rho_1^{AB}/\partial t = -iJ[\hat{S}_A\hat{S}_B, \rho_1^{AB}] - (1 - f)\rho_1^{AB}/\tau_0 + f\rho_2^{AB}/\tau_0 ,$$
$$\partial\rho_2^{AB}/\partial t = (1 - f)\rho_1^{AB}/\tau_0 - f\rho_2^{AB}/\tau_0 . \tag{2.187}$$

The solution of these equations may be expressed through the linear operator $\hat{Q}$

$$\rho^{AB}(t) = \hat{Q}(t)\rho^{AB}(0) . \tag{2.188}$$

The operator $\hat{Q}(t)$ is averaged according to (2.179) with the help of the distribution function of the form of (2.115) with the average time τ_c' of one contact. For the model described, in [2.83] the averaged operators F and Q have been found. Substituting them into (2.182) one determines the efficiency of spin exchange which may be expressed in the form of (2.171). The parameter χ, however, now has a different meaning; it reflects the averaging of spin exchange anisotropy both at the moment of contact and in the intervals between contacts

$$\chi = (1 + qn + \tau_c/\tau_o)/[1 + f(qn + \tau_c/\tau_o)] \quad . \tag{2.189}$$

Here $n = 1/(1 - \bar{p})$ is the number of all contacts of a given pair, $\tau_c = n\tau_c'$ is the total time of all the contacts, and q characterizes the efficiency of orientational relaxation in the interval between two successive contacts

$$q = \int f(t)[1 - \exp(-t/\tau_o')]dt = \bar{p} - \bar{p} \cdot \exp(-m/\sqrt{\tau_o'}) \quad . \tag{2.190}$$

Spin exchange occurs in a certain narrow layer of interaction. Therefore τ_c is essentially the total lifetime of two partners in this layer having a thickness a. If the diffusion of particles proceeds via long jumps, i.e., $\lambda \gg a$, then the lifetime of one contact τ_c' is given by (2.118) and the number of all contacts may be estimated with the help of the relation $n = 1/(1 - \bar{p}) \cong (R_{AB} + \lambda)/\lambda$ [2.26]. In the extreme model of continuous diffusion the total litetime of the partners in the interaction layer is given by (2.121). Equation (2.121) appears to yield a satisfactory estimation of τ_c for the jump model of diffusion in the case $\lambda \leq a$ as well [2.26].

Orientational relaxation manifests itself differently for weak and strong exchange. In the case of weak exchange according to (2.171)

$$p \cong fJ^2\tau_c^2/2\chi \quad . \tag{2.191}$$

Averaging the anisotropy by orientational relaxation decreases the efficiency of spin exchange χ times compared to its value without this relaxation. In the opposite case of strong exchange according to (2.171)

$$p \cong f_e/2 = f\chi/2 \quad . \tag{2.192}$$

Due to the averaging of anisotropy the efficiency of spin exchange increases by χ times.

In order to quantitatively estimate the influence of orientational relaxation one should choose a definite model for the diffsuion of particles. The simplest results are obtained for a model of continuous diffusion. In this case τ_c is given by (2.121), τ_0 may be estimated in terms of (2.174), and qn is equal to [2.26]

$$qn \Rightarrow [R_{AB}^2/D_{AB} \cdot \tau_0']^{\frac{1}{2}} \quad .$$

Substituting these quantities into (2.189) we obtain at $\tau_0' = \tau_0$

$$\chi = (1 + \sqrt{6} + 6a/R_{AB})/[1 + f(\sqrt{6} + 6a/R_{AB})] \quad . \qquad (2.193)$$

From (2.193) it follows that for partners with pronounced anisotropy of interaction when $f \leq 0.1$, $\chi \cong 4$, i.e., the effective steric factor exceeds by about 4 times the geometric steric factor f. Thus the rotation of the partners averages essentially the exchange anisotropy. (The particle translational motion between reencounters seems contribute appreciably to the averaging effect, see, e.g., R. Samson and J.M. Deutsh. J. Chem. Phys. *68*, 285 (1978). Phenomenologically it means τ_0 must be less than that calculated by (2.174). Notes added in proof).

2.5.5 Spin Exchange of Radicals with Other Paramagnetic Particles

In the theory of spin exchange in binary solutions containing along with free radicals (particles A) some other paramagnetic species (particles B) - paramagnetic complexes of transition metals, triplet excited molecules, etc. - a number of new problems arise which were practically of no significance in the theory of spin exchange between radicals. The partner B may have an arbitrary spin, so it is necessary to analyze the dependence of exchange rate on the value of S_B. In the case of exchange between radicals at the moment of contact one could neglect the comparatively weak intramolecule spin interactions: hfi with magnetic nuclei, spin-orbital interaction. Paramagnetic complexes and triplet molecules may have sufficiently large intrinsic spin energies (spin-orbital and spin-spin dipole-dipole interaction between unpaired electrons when $S_B > 1/2$). In this case these intrinsic interactions of S_B should be taken into account at the moment of collision of particles along with the exchange interaction between them.

It is convenient to introduce a parameter D which characterizes the strength of intrinsic spin interactions of S_B. Let τ_0 be the correlation time for random modulation of D by the oscillations of atoms or by the rotation of a paramagnetic particle. If D does not take too large values so that $\langle D^2 \rangle \tau_c^2 < 1$ then, just as in the case of radicals, the internal spin interactions of unpaired electrons play a negligible part in the process of spin exchange. But in practice it may also occur that $<D^2>\tau_c^2 \gtrsim 1$. Then the picture of spin exchange and the character of the interference of exchange interaction with D depend on the extent to which the thermal motion of molecules is able, within the contact lifetime of colliding spins, to average D. If $\tau_0 > \tau_c$, then within the time of collision D conserves a certain fixed value. In this case D must reveal itself in spin exchange qualitatively in just the same way as the difference of the Zeeman frequencies of spin partners at $\delta\tau_c = (\omega_A - \omega_B)\tau_c > 1$ whose role was discussed in the foregoing subsections. The results obtained there confirm that at $D\tau_c > 1$, $\tau_0 > \tau_c$ the exchange interaction brings about a dephasing of spins rather than an equivalent exchange of spin states between the partners. This problem has not been studied in detail so far. If $\tau_0 < \tau_c$ then within the time of collision D is effectively averaged, and the role of D in the exchange process can be taken into account by introducing the corresponding relaxation terms into the equation for the tensor of collision efficiency.

Spin exchange between radicals and spins S_B was theoretically studied in the two limiting cases which appear to cover the practically most important situations. On the one hand, analysis is made of the spin dependence of the exchange rate in the systems when at the moment of collision the influence of D on the motion of the spins is negligible [2.72-74]. This group of systems includes most of the real systems of spins S_B with long times of paramagnetic relaxation, T_{1B}, $T_{2B} > \tau_c$. On the other hand, a study has been made of the spin exchange for the situation of short times of paramagnetic relaxation, T_{1B}, $T_{2B} < \tau_c$ [2.77,78]. At present, however, there is no theory of spin exchange between radicals and particles with arbitrary spins S_B for the extreme situations $D\tau_c > 1$, $\tau_0 > \tau_c$ or $<D^2>\tau_0^2 > 1$, $\tau_0 < \tau_c$.

Spin Exchange Between Radicals and Spins S_B with Long Times of Paramagnetic Relaxation

In this case one may neglect the influence of internal spin interactions on the elementary act of spin exchange and leave in the spin-Hamiltonian of the pair of spins of A and B, which are in contact, only their Zeeman interaction with the external field and their exchange interaction with each other. The problem in such a formulation is a direct generalization of the problem of spin exchange between two radicals with different Zeeman frequences that was considered in the previous subsection. The only difference is that now the value of S_B is arbitrary. The problem formulated was solved in [2.72-74].

The kinetic equations for the density matrix of the radical and the spin S_B are given in (2.133) and the tensor of collision efficiency is determined by (2.135). Linearizing (2.133) and introducing the macroscopic magnetizations of the spins of A (radicals) and B, from (2.133,135) one may find the necessary kinetic equations. Thus, e.g., in the most interesting situation of equivalent spin exchange, the following results are obtained. The kinetic equations for the components of magnetization take the form [2.78]

$$\begin{aligned}
\partial M_-^A/\partial t &= i\omega_A M_-^A - K_{eA}C_B M_-^A + K_{eB}C_A M_-^B - M_-^A/T_{2A} \ , \\
\partial M_-^B/\partial t &= i\omega_B M_-^B + K_{eA}C_B M_-^A - K_{eB}C_A M_-^B - M_-^B/T_{2B} \ , \\
\partial M_z^A/\partial t &= -K_{eA}C_B M_z^A + K_{eB}C_A M_z^B - (M_z^A - M_0^A)/T_{1A} \ , \\
\partial M_z^B/\partial t &= K_{eA}C_B M_z^A - K_{eB}C_A M_z^B - (M_z^B - M_0^B)/T_{1B}
\end{aligned} \qquad (2.194)$$

and the exchange rate constants for the spins of A and B prove to be equal to

$$\begin{aligned}
&K_{eA} = p_A K_D \ , \quad K_{eB} = p_B K_D \ , \\
&p_A = (2/3)S_B(S_B + 1)J^2\tau_c^2/[1 + J^2\tau_c^2(S_B + 1/2)^2] \ , \\
&p_B = (1/2)J^2\tau_c^2/[1 + J^2\tau_c^2(S_B + 1/2)^2] \ .
\end{aligned} \qquad (2.195)$$

As seen from the above results, the exchange efficiency depends strongly on the magnitude of the spin of the partners. The same collisions change the spin states of the partners differently: the large spin is much more efficient in reorientation of the small spin than is the small spin in reorientation of the large one. This result is illustrated in Fig.2.7. The change of the spins' orientation induced by exchange interaction proves to be due to their precession around the total spin moment of the partners. In this case from the precession model it follows that the spin $S_B > 1/2$, even at comparatively long times τ_c, changes its orientation in space less than does the spin $S_A = 1/2$ of the partner.

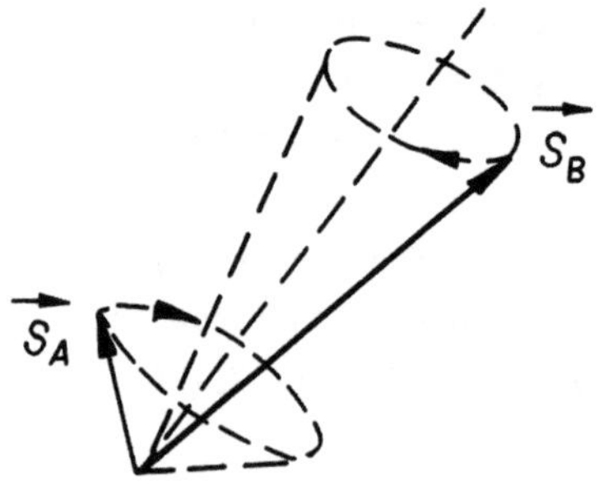

Fig. 2.7. The scheme of spin precession explaining the influence of the partners' spin values on the efficiency of spin exchange during their collision. Both spins precess around the direction of the vector of the total spin. The variation of the orientation is smaller for the bigger spin S_B

According to (2.195) the ratio of two exchange rates constants does not depend on J, and is equal to

$$K_{eA}/K_{eB} = (4/3)S_B(S_B + 1) = S_B(S_B + 1)/S_A(S_A + 1) \quad . \qquad (2.196)$$

It is of interest to note that the exchange efficiency, even for strong exchange $J\tau_c > 1$, does not reach unity. This is due to the fact that in the problem under consideration the projections of the total spin of both partners are conserved in the process of a collision. From (2.195) at $J\tau_c > 1$ we have

$$p_A(\max) = (4/3)S_B(S_B + 1)/(S_B + 1/2)^2 \quad ,$$
$$p_B(\max) = (1/2)/(S_B + 1/2)^2 \quad . \qquad (2.197)$$

Shape of ESR Spectra

We shall consider briefly some peculiarities of the spectral manifestations of spin exchange in solutions of particles with different spins.

Kinetic equations (2.194) coincide with the equations for $\underline{M}^A$ and $\underline{M}^B$ in the chemical exchange of a particle between two nonequivalent sites with different frequencies ω_A and ω_B and the statistical weights of the states of A and B equal to $\tau_A/(\tau_A + \tau_B)$ and $\tau_B/(\tau_A + \tau_B)$, respectively, where $\tau_A^{-1} = K_{eA}C_B$, $\tau_B^{-1} = K_{eB}C_A$ [2.35].

Qualitatively, the spin exchange between radicals and spins S_B in the ESR spectra of their solutions manifests itself in just the same way as does the spin exchange between radicals discussed in detail in Sect. 2.4. There may be extreme situations of slow and fast exchange yielding, respectively, a broadening of the lines of the radical and of its partner, and an exchange narrowing of the entire ESR spectrum. There is some difference, however, in the criterion of slow and fast exchange. Radicals are characterized by long times T_{2A} and for them the splitting of the ESR spectra $\langle\Delta\omega_A^2\rangle^{1/2}$ is large compared to $1/T_{2A}$, i.e., $T_{2A} \cdot \langle\Delta\omega_A^2\rangle^{1/2} > 1$. Therefore the criterion of slow and fast exchange is determined only by the relation between $\langle\Delta\omega^2\rangle^{1/2}$ and the exchange rate (2.68,69). For arbitary spins S_B even in the case under discussion $(T_{2B} > \tau_c)$ T_{2B} may be a relatively small quantity of the order of 10^{-8}-10^{-10} s so that the relationship $T_{2B} \cdot |\omega_A - \omega_B| < 1$ may be fulfilled. Then the criteria of slow and fast exchange (2.68) and (2.69) cease to be valid.

A detailed analysis of the spectral manifestations of the exchange between two frequencies made in [2.38] shows that at

$$|\omega_A - \omega_B| > \max\{K_{eA}C_B, K_{eB}C_A\} \qquad (2.198)$$

or

$$|\omega_A - \omega_B| < \max\{K_{eA}C_B, K_{eB}C_A\} < 1/T_{2B} \tag{2.199}$$

the spectrum lines of the radical and its partner are broadened, in fact, independently, and the exchange broadening is given by the equations

$$\Delta\omega_{eA} = K_{eA}C_B \;, \quad \Delta\omega_{eB} = K_{eB}C_A \quad . \tag{2.200}$$

From (2.199) it is seen that for sufficiently short times T_{2B} when $|\omega_A - \omega_B|^{-1} > T_{2B} > \tau_c$ the spectrum does not collapse despite the fulfilment of the usual condition of fast exchange

$$|\omega_A - \omega_B| < \max\{K_{eA}C_B, K_{eB}C_A\} \quad . \tag{2.201}$$

This is because in the interval between effective collisions, $t \sim 1/K_eC$, the spins S_B have time to relax completely and so between successive collisions no phase memory remains.

In the case of comparatively short times T_{2B} when $T_{2B} \cdot |\omega_A - \omega_B| < 1$ the condition of exchange narrowing of the spectrum takes the form

$$\max\{K_{eA}C_B, K_{eB}C_A\} > 1/T_{2B} \quad . \tag{2.202}$$

Note one interesting but experimentally unstudied peculiarity of the manifestation of spin exchange in which particles with a spin $S_B > 1$ take part. In liquid solutions with this kind of particles the fine structure of the ESR spectrum does not manifest itself, as a rule, and the different transitions between equidistant Zeeman levels yield the absorption line at one and the same frequency. This line is broadened by spin exchange with a radical at the efficiency p_B (2.195). If conditions were created allowing the fine structure of the ESR spectrum of S_B to be observed then the components of the ESR spectrum of S_B will be broadened by spin exchange differently. For instance, if $S_B = 5/2$, then the lines corresponding to transitions $-5/2 \Leftrightarrow -3/2$ and $3/2 \Leftrightarrow 5/2$ will be broadened by exchange with a radical with the efficiency $p_B' = 5p_B = 5J^2\tau_c^2/2(1 + 9J^2\tau_c^2)$, the lines on the transitions $1/2 \Leftrightarrow 3/2$ and $-3/2 \Leftrightarrow -1/2$, with the efficiency $p_B'' = 8p_B$, and the line $-1/2 \Leftrightarrow 1/2$ with the efficiency $p_B''' = 9p_B$ [2.23]. It is interesting that the lines of individual transitions are broadened by exchange essentially more than the unresolved line which is simultaneously contributed to by all transitions between the equidistant Zeeman levels of S_B. In the latter case the line of S_B is broadened with the efficiency $p_B = (\frac{1}{2})J^2\tau_c^2/(1 + 9J^2\tau_c^2)$. The effect in question can probably be observed, e.g., for exchange between mobile electrons or triplet excitons and fixed high-spin paramagnetic centers in crystals.

The Influence of the Relaxation of Spin S_B on the Rate of Exchange

Consider the exchange between radicals and spins S_B, which have such short times of their intrinsic paramagnetic relaxation

$$T_{1B}\ ,\ T_{2B} < \tau_c \qquad (2.203)$$

that no resonance absorption by such spins S_B in the fields $H_0 \simeq 3000$ G is observed. Examples of such spins may be the paramagnetic ions Co^{2+}, Fe^{2+}, Ni^{2+} for which in solutions T_{1B}, $T_{2B} \sim 10^{-12}$-10^{-13} s. Since in such systems one observes the ESR absorption lines only for the radical, the spin exchange manifests itself only in shortening the times of the longitudinal and transverse relaxation of radical spins, i.e., in a broadening of the ESR lines of the radical. As a matter of fact, the spins S_B with short relaxation times serve for radicals as effective acceptors of spin excitation, so with respect to radicals the spins S_B can be considered as a part of the lattice thermostat.

The fast relaxation of the spin S_B influences significantly the process of its exchange with the radical. This influence arises for two reasons. Firstly, the relaxation of spin S_B varies randomly the exchange field in which the spin of the radical is moving. The effect of the averaging of exchange interaction by the relaxation of the spin S_B is similar to that of averaging by the rotation of anisotropic exchange interaction. Secondly, if (2.203) is fulfilled, then the projections of the total spin of the colliding partners are no longer constants of motion, so the spin state of the radical may also change arbitrarily in the collision. As a result, the exchange efficiency for the radical may reach 1 [cf with (2.197)].

As before one can differentiate the situations of weak and strong exchange. In the case of weak exchange when

$$JT_{1B}\ ,\ JT_{2B} < 1 \qquad (2.204)$$

the spin excitation of the radical is transferred in the course of collision to the spin S_B and then quickly relaxes. The limiting step for the process of changing the spin state of the radical is the exchange with the partner. In the opposite case of strong exchange when

$$JT_{1B}\ ,\ JT_{2B} > 1 \qquad (2.205)$$

the exchange between A and B proceeds rapidly so that the collision partners relax as a united exchange-bound pair. As a result at the moment of collision the spin S_A is characterized by the same times of paramagnetic relaxations as is the partner. Keeping in mind, also, that (2.203) holds here, we arrive at the conclusion that in the case of (2.205) within the time of collision the spin of the radical has time to relax and the exchange efficiency p_A for the radical must equal 1.

As has been noted earlier, in the system under discussion the spins S_B serve for the spins S_A as a thermostat. This allows one to consider the collision between S_A and S_B in the approximation of the external field assuming S_B to create a definite exchange potential in which the spin S_A is scattered. This exchange field is believed to vary randomly in time due to the relaxation of S_B. The exchange interaction in the case of collision between the radical and the spin S_B with short times of paramagnetic relaxation may be written in the form

$$\hat{V}(t) = \hbar J\hat{S}_B(t)\hat{S}_A \quad . \tag{2.206}$$

Considering in the quasiclassical approximation $J\hat{S}_B(t)$ as a random parameter we find that for the operator of collision efficiency (2.159,160) remain formally valid. These equations were obtained when discussing the random modulation of anisotropic exchange interaction by molecular rotation

$$\begin{aligned} &(\partial/\partial t)\langle\hat{T}(t)\rangle = -[\langle\hat{T}(t)\rangle, \hat{Q}] + \langle\hat{\Omega}(t)\hat{T}(t)\rangle \ , \\ &\Omega_{mn,k\ell}(t) = -iJ[(S_B(t)S_A)_{mk}\delta_{n\ell} - (S_B(t)S_A)_{\ell n}\delta_{mk}] \ , \end{aligned} \tag{2.207}$$

where <...> stands for averaging over all realizations of $\underline{S}_B(t)$ and over the duration of collisions.

It is comparatively simple to perform the averaging in (2.207) in the case of weak exchange when one can apply the stochastic theory of perturbations on $\hat{V}(t)$ [2.31]. In the second order of this theory for the model of sudden collisions from (2.207), e.g., in the case of $T_{1B} = T_{2B} = T_B$, we obtain [2.23]

$$[\hat{T}, \hat{Q}_A] = -(\hat{T} - \hat{E})/\tau_c - J^2T_B[\hat{S}_A\hat{S}_B(0), [\hat{S}_A\hat{S}_B(0), \hat{T} - \hat{T}_0]] \ , \tag{2.208}$$

where we introduce an operator $\hat{T}_0$ which transfers the two-particle density matrix to its equilibrium value

$$\hat{T}_0\rho^{AB} = \rho_0^A \times \rho_0^B \quad .$$

Solving the system of algebraic equations (2.208) for the elements of $\hat{T}$, substituting the results obtained into the kinetic equation for the one-particle density matrix of the spin S_A (2.131), and bearing in mind that due to the fast relaxation of S_B in the right-hand side of (2.131) ρ^B can be replaced by its equilibrium value, i.e., one can assume $\rho^A \times \rho^B \cong \rho^A \times \rho_0^B$, we come to the following kinetic equation:

$$(\partial/\partial t)\rho^A_{\frac{1}{2},-\frac{1}{2}} = -i\omega_A\rho^A_{\frac{1}{2},-\frac{1}{2}} - K_{eA}C_B\rho^A_{\frac{1}{2},-\frac{1}{2}} , \tag{2.209}$$

where the exchange rate constant K_{eA} is equal to

$$K_{eA} = (2/3)S_B(S_B + 1)J^2T_B\tau_cK_D \quad . \tag{2.210}$$

In a more general case when $T_{1B} \neq T_{2B}$ a similar consideration gives

$$K_{eA} = (1/3)S_B(S_B + 1)J^2(T_{1B} + T_{2B})\tau_cK_D \quad . \tag{2.211}$$

Comparison of (2.211) with (2.195) shows that in the case of weak exchange the relaxation of the partner decreases the exchange efficiency by a factor $(\tau_c/T_B) > 1$ due to the averaging action of relaxation. This effect can be as large as one or two orders of magnitude for such ions as Co^{2+}, Fe^{2+}, and Ni^{2+}. This decrease in efficiency can be interpreted in terms of a precession model. The orientation of one of the spins within the time of collision randomly changes and, hence, so does the spatial direction of the axis about which the other spin precesses. The motion of the tip of the vector of this spin over the sphere will remind one of the Brownian motion with a diffusion coefficient $\sim J^2T_B$, and this will lead to far smaller deviations of the spin from the initial state than will the regular precession.

In principle (2.208) allows us to go beyond the framework of the perturbation theory and take into account the influence of spin relaxation on the exchange at arbitrary J, T_{1B}, T_{2B}. This problem was solved [2.88] by assuming $S_B = 1/2$. The calculations made confirmed the conclusion that in the general case the exchange constant for a radical may be represented in the form suggested in [2.77,78]

$$K_{eA} = p_AK_D \ , \quad p_A = k_e\tau_c/(1 + k_e\tau_c) \ ,$$
$$k_e = (1/3)S_B(S_B + 1)J^2(T_{1B} + T_{2B}) \quad . \tag{2.212}$$

Hence it follows, as expected, that in the case of strong exchange the spin dependence of the exchange efficiency vanishes and $K_{eA} \rightarrow K_D$.

It is to be noted that the random modulation of exchange interaction by partner relaxation has much in common with the influence of rotation on spin exchange. However, there are also some qualitative differences between the two effects: the modulation of anisotropic interaction by the rotational diffusion of molecules, even in the case of strong exchange, does not destroy the spin dependence of the exchange rate, and $1/2 \leq p_A \leq 2/3$, as distinct from the influence of relaxation when according to (2.212) p_A no longer depends on S_B and $p_A = 1$ when the exchange becomes strong.

Anisotropy of Spin Exchange

The theory of anisotropic spin exchange between particles with $S = 1/2$ was also generalized to solutions containing two kinds of particles with the spins $S_A = 1/2$ and $S_B > 1/2$ [2.83]. The efficiency of exchange between a radical ($S_A = 1/2$) and paramagnetic particles with sufficiently long times of relaxation, T_{1B}, $T_{2B} > \tau_c$ is [cf with (2.171,195)]

$$p_A = (2/3)S_B(S_B + 1)f_e J_e^2 \tau_c^2/[1 + J_e^2 \tau_c^2 (S_B + 1/2)^2] \tag{2.213}$$

and the efficiency of spin exchange for B is

$$p_B = (1/2)f_e J_e^2 \tau_c^2/[1 + J_e^2 \tau_c^2 (S_B + 1/2)^2] \ , \tag{2.214}$$

where f_e and J_e are given by (2.171).

For the interaction between a radical and paramagnetic particles with relatively short times of relaxation, T_{1B}, $T_{2B} < \tau_c$, qualitative changes in the dynamics of an elementary act of spin exchange occur: dephasing of radical precession rather than an equivalent exchange of spin projections between the partners dominates. In this case at the moment of contact the phase relaxation of the radical formally coincides with the kinetics of recombination of two contacting radicals [2.26,85]. Therefore the orientational relaxation of the partners averages the anisotropy of spin exchange in just the same way as it does that of reactivity in radical recombination [2.85]. The efficiency of anisotropic spin exchange (dephasing) between radicals and particles with short times of paramagnetic relaxation is given by [2.83]

$$\begin{aligned} p_A &= f_e k_e \tau_c/(1 + k_e \tau_c) \ , \quad f_e = f\chi \ , \\ k_e &= (1/3)S_B(S_B + 1)J^2(T_{1B} + T_{2B})/\chi \ , \end{aligned} \tag{2.215}$$

where χ is given by (2.189).

From the above it follows that anisotropy averaging always increases the efficiency of the process χ times in the case of strong exchange, and decreases it as much in the case of weak exchange for partners with long times of paramagnetic relaxation, but leaves it practically unchanged for those with short ones.

2.5.6 Spin Exchange in Biradicals

Biradicals are species with two unpaired electrons localized on different atoms. From the point of view of spin exchange biradicals are interesting in many respects. In a series of biradicals one may realize various situations for exchange interaction. When unpaired electrons are localized close to one another a situation is created in a biradical that can model an exchange interaction between paramagnetic particles in magnetically concentrated liquids or solids.

When the unpaired electrons of a biradical are separated by a great number of chemical bonds the exchange interaction in the biradical simulates the situation in magnetically diluted solutions.

A biradical may occur in a variety of conformations. Exchange interaction in a biradical depends on its conformation. Transitions between the conformations induce random variations of the exchange integral. So the dynamics of spins in a biradical coincides, in effect, with that of spins in a pair of particles that collide in a solution when, say, the anisotropic exchange integral is randomly modulated by rotational diffusion. Therefore for the spin density matrix in a biradical the equations considered in Sects.2.5.2-4 for dimers - spin pairs in a cage - remain valid.

Depending on the total number of conformations in a biradical, on the value of the exchange integral for individual conformations, and on the rate of transitions among various conformations there may be several extreme situations of spin dynamics in biradicals [2.89]. If transitions between conformations occur very seldom, then there is just a set of biradicals each of which is characterized by its intrinsic absorption spectrum and during an experiment we shall observe a superposition of ESR spectra of different biradicals. The reduction of the lifetime τ_e of a biradical in one or another conformation may broaden the components of their ESR spectrum by the value of $\Delta\omega \sim 1/\tau_e$. Note that the situation here is quite similar to the changes in the nuclear magnetic resonance spectra that occur as a result of a random modulation of the j-j interaction between nuclear spins by a process of chemical exchange among various conformations of molecules [2.32]. For the case of a high intramolecular mobility an analysis of the ESR spectra of biradicals was made [2.90]. Nitroxide biradicals with a sufficiently large mean value of the exchange integral were studied. In this case the modulation of exchange interaction was shown to broaden the different ESR lines differently. A number of interesting models of chemical exchange of a biradical between two and three conformations were considered in [2.91,92] where simulation programs of the ESR spectra in a biradical were given for an arbitrary rate of intramolecular motion.

Spectral effects of exchange interaction in biradicals are fully enough reflected in a number of reviews [2.22,75]. Note that in the literature on biradicals the term spin exchange is often used to cover any manifestations of exchange interaction, including those in biradicals with a frozen configuration and strong interaction $\hbar J > kT$ when it is already necessary to introduce collective spin states of a biradical as a whole. In this section we shall discuss only those aspects that are closely connected with the problem of spin exchange during short-time pair collisions of paramagnetic particles, i.e., we shall discuss biradicals for which it proves appropriate to speak about a process of spin exchange between the radical centers in the sense this term was defined in the introduction to this book.

The Model

Assume two radical centers of a biradical to be bound by a sufficiently long and flexible chain of chemical bonds. As a result of heat motion of this chain, from time to time the radical centers will draw closer to each other and an exchange interaction will then ensue. If the radical centers are separated by a sufficiently long chain of the order of 10 or more chemical bonds, then the time τ_c of such a contact will be much shorter than the time τ_0 between contacts.

In fact, we come to a model of pair collisions of radicals in diluted solutions. The only difference is that in a biradical repeated collisions of one and the same pair of radicals occur while in a solution a radical collides with different partners. Thus within the model chosen the biradical remains mostly in a state in which the exchange interaction between the radical centers is negligible (we shall denote it as state I) and only for relatively short intervals of time is the exchange interaction switched on (state II). Within this time a spin exchange occurs between two paramagnetic centers of a biradical.[5]

---

5 It should be emphasized that the present model is far from exhausting the situations that can be realized in biradicals. For example in reality many biradicals seem to have $\tau_c \approx \tau_0$ [2.89,91]. In these cases the effects produced by the exchange interaction of the paramagnetic centers of a biradical cannot be analyzed using the collision model under discussion. One has to turn to the general methods developed in the theory of chemical exchange between different states which are characterized by different values of the spin-Hamiltonian parameters [2.32,37,93,94]. Practical recommendations for measuring the spin exchange characteristics at $\tau_c \sim \tau_0$ can be found in [2.89,91].

The model discussed appears to be applicable for polymer molecules with paramagnetic labels [2.95]. The study of the process of spin exchange between labels may yield information on the intramolecular mobility of macromolecules, among them biopolymers, in solutions.

Kinetic Equations

The equations for the spin density matrix of a biradical performing transitions between conformations with different values of the exchange integral J has the form

$$\partial\rho/\partial t = -i\hbar^{-1}[\hat{H}_0 + \hat{V}(t), \rho] + (\partial\rho/\partial t)_{relax.} , \qquad (2.216)$$

where the last term describes the change of ρ due to the relaxation mechanisms unrelated to exchange interaction; $\hat{H}_0$ is the Zeeman energy and the isotropic hfi of unpaired electrons with magnetic nuclei; and $\hat{V}(t) = \hbar J(t)\hat{S}_1\hat{S}_2$ is the spin-Hamiltonian of exchange interaction. Equation (2.216) must be averaged over all possibilities of random changes in the exchange integral J(t). The technique of such an averaging was considered above (e.g., in discussing the influence of radical rotation on spin exchange). For the chosen model of transitions between the two states I and II the averaging (2.216) leads to the following kinetic equations for partial density matrices ρ^I and ρ^{II} of the subensembles of biradicals in states I and II [2.32,89]:

$$\begin{aligned} \partial\rho^I/\partial t &= -i\hbar^{-1}[\hat{H}_I, \rho^I] - (\rho^I - \rho^{II})/\tau_0 + (\partial\rho^I/\partial t)_{relax.} , \\ \partial\rho^{II}/\partial t &= -i\hbar^{-1}[\hat{H}_{II}, \rho^{II}] + (\rho^I - \rho^{II})/\tau_c , \end{aligned} \qquad (2.217)$$

where $\hat{H}_I = \hat{H}_0$, $\hat{H}_{II} = \hat{H}_0 + \hbar J\hat{S}_1\hat{S}_2$. On the basis of the assumption that τ_c is much shorter than the paramagnetic relaxation times of radicals and that the change in the spin density matrix during the radicals contact time is negligible, in the equation for ρ^{II} (2.217) the relaxation term $(\partial\rho^{II}/\partial t)_{relax.}$ is omitted. A generalization of (2.217) for the cases when the exchange integral takes a variety of values was first made by JOHNSON [2.93]. These results were further developed in quite a number of studies [2.23,37,94].

Just as in the sudden collisions model we passed from the Lynden-Bell equations (2.125) to the equations in an impact approximation (2.131), so in (2.217), too, one can make the following simplifications. Solving the equation for ρ^{II} we obtain

$$\rho^{II}(t) = \exp(-t/\tau_c) \cdot \exp(-i\hbar^{-1}\hat{H}_{II}t)\rho^{II}(0) \times \exp(i\hbar^{-1}\hat{H}_{II}t)$$
$$+ \int (d\tau/\tau_c) \exp(-\tau/\tau_c)$$
$$\times \exp(-i\hbar^{-1}\hat{H}_{II}\tau)\rho^{I}(t - \tau) \exp(i\hbar^{-1}\hat{H}_{II}\tau) \quad . \tag{2.218}$$

The changes of ρ^I for a short time interval τ_c are only due to the action of the Hamiltonian $\hat{H}_I$. Therefore in (2.218) one can assume

$$\rho^I(t - \tau) \cong \exp(i\hbar^{-1}\hat{H}_I\tau)\rho^I(t) \exp(-i\hbar^{-1}\hat{H}_I\tau) \quad . \tag{2.219}$$

Substituting (2.218) and (2.219) into the equation for ρ^I (2.217) and bearing in mind that for $t \sim \tau_0$ $\exp(-t/\tau_c) \to 0$ we obtain the following kinetic equation for biradical

$$\partial\rho/\partial t = -i\hbar^{-1}[\hat{H}_0, \rho] - (\rho - \hat{T}\rho)/\tau_0 + (\partial\rho/\partial t)_{relax.} \ ,$$
$$[\rho = \rho^I\tau_0/(\tau_0 + \tau_c) + \rho^{II}\tau_c/(\tau_0 + \tau_c) \cong \rho^I] \ , \tag{2.220}$$

where the efficiency tensor of spin exchange in a biradical, as has been expected, is given by the same relation (2.132) that was true for a collision of two radicals in a solution. Hence it follows that the rate of spin exchange in a biradical is equal [2.28] to

$$\omega_e = (1/2\tau_0) \cdot J^2\tau_c^2/(1 + J^2\tau_c^2) \quad . \tag{2.221}$$

The ESR Spectrum of Nitroxide Biradicals

Nitroxide radicals are especially often used as spin labels in experiments with biopolymers. We shall therefore consider the spectral effects of spin exchange on an example of a biradical containing two nitroxide radical centers.

The spin-Hamiltonian of a radical includes the Zeeman energy and hfi with a nitrogen nucleus

$$\hat{H}_0 = \hbar\omega_0\hat{S}_z + \hbar a\hat{I}_z\hat{S}_z \quad . \tag{2.222}$$

The projections of a nuclear spin I of nitrogen (^{14}N) on the direction of the external magnetic field that are equal to $m = 0, \pm 1$, correspond to three components, equal in intensity, of the ESR spectrum of a radical with a splitting between the components that equals a hfi constant (Fig.2.8a). The biradicals will have the following combinations of nitrogen nuclear spin projections (m_1, m_2):

a) (+1, +1), (0,0), (-1,-1);
b) (+1,0), (0,+1), (0,-1), (-1,0);
c) (+1,-1), (-1,+1).

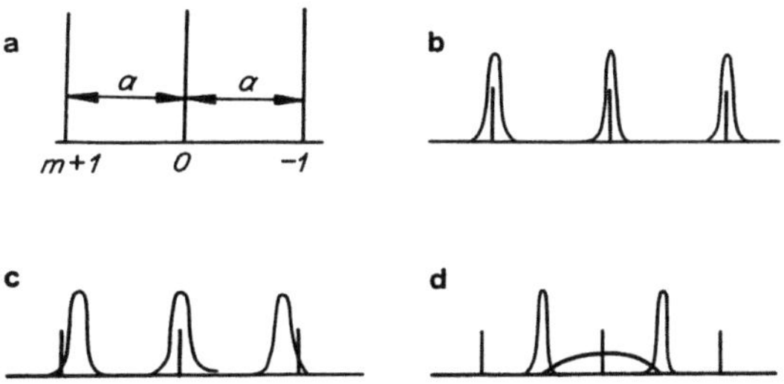

Fig. 2.8a-d. ESR spectrum of a nitroxide biradical. (a) Spectrum of a monoradical; (b,c) slow spin exchange; (d) fast spin exchange

In biradicals with the same spin orientations of both nuclei (case a) the spin exchange must not reveal itself in the ESR spectra since it does not induce any frequency exchange. Therefore the subensemble of biradicals corresponding to the case a will have a ESR spectrum coinciding with that of separate radicals (see Fig.2.8a). In the biradicals with different orientations of nuclear spins (cases b and c) the spin exchange leads to a frequency exchange and will therefore transform the ESR spectrum: in the region of slow exchange, when $\omega_e < a$, the spectrum components will get broader by the quantity

$$\Delta\omega_e = \omega_e = (1/2\tau_0)J^2\tau_c^2/(1 + J^2\tau_c^2) \qquad (2.223)$$

and in the region of fast exchange, when $\omega_e > a$, the two spectrum components merge into one narrow line. The spin exchange makes a contribution to the width of this collapsed line equal to

$$\Delta\omega_e = a^2(m_1 - m_2)^2/8\omega_e = a^2(m_1 - m_2)^2\tau_0(1 + J^2\tau_c^2)/4J^2\tau_c^2 \quad . \qquad (2.224)$$

The observed spectrum respresents an additive sum of spectra of biradical subensembles with different configurations of nuclear spins of nitrogens. In the slow exchange region the spectrum of a biradical consists, just as that of a nitroxide monoradical, of three components. Each component, however, has rather a complex form. It represents a sum of two lines, $g_1(\omega)$ and $g_2(\omega)$, with the ratio of intensities 1:2; $g_1(\omega)$ corresponds to the ESR spectrum component of a monoradical and has a width $1/T_2$ while $g_2(\omega)$ is a line which is exchange broadened and shifted toward the spectrum center. The width of $g_2(\omega)$ equals $\omega_e + 1/T_2$ (see Fig.2.8b). As the exchange rate grows the outer lines of the spectrum split into two components from $g_1(\omega)$ and $g_2(\omega)$ due to a shift of the line $g_2(\omega)$ (see Fig.2.8c). And, lastly, in the region of fast exchange the spectrum of the biradical will contain five components with the ratio of intensities 1:2:3:2:1 and splitting between the components equal to a/2 (see Fig. 2.8d). The two outer spectral components are not perturbed by exchange and have the shape of g_1 with the width $1/T_2$. The central component is a sum of two contributions: one is from biradicals with the configuration of nuclear nitrogen spins (0,0); it is not perturbed by exchange and has the width $1/T_2$. The other contribution which is twice as intensive stems from biradicals with the configuration of nuclear spins (+1,-1),(-1,1) and has the width $1/T_2 + a^2/2\omega_e$. The second component corresponds to biradicals with the nuclear spin configurations (0,1), (1,0) and the 4th component corresponds to configurations (0,-1), (-1,0). Both components have the width $1/T_2 + a^2/8\omega_e$ (Table 2.5). It is of interest to note that in the case of fast spin exchange the broadening of separate ESR spectrum components of a biradical turn out to be of the same character as in the situation studied in [2.90], but quantitatively the contribution to the line width in the spin exchange model under discussion is different from the results obtained in the work just mentioned. Thus we see that a detailed analysis of spin exchange manifestations in ESR spectra of biradicals allows one to find ω_e, i.e., may yield information on the intramolecular mobility of long-chain molecules.

Table 2.5. Contribution of spin exchange to the broadening of ESR lines of biradicals with different configuration of nuclear spins

Configuration of nitrogen nuclear spins (m_1, m_2)	Slow exchange		Fast exchange	
	Line position	Exchange contribution to line width	Line position	Exchange contribution to line width
±1 ±1	±a	0	±a	0
0 ±1				
±1 0	0, ±a	ω_e	±a/2	$a^2/8\omega_e$
0 0	0	0	0	0
±1 ∓1	∓a	ω_e	0	$a^2/2\omega_e$

2.6 The Model of Diffusive Passage

As has been mentioned above, in pair collisions of particles in solutions the cross section of the processes is determined both by the interaction potential and by the kinematics of passing the region of interaction by the particles [2.96]. The preceding section dealt with the results of the spin exchange theory obtained in the extreme case of passing the region of interaction in one jump. Such an approximation, however, is unsatisfactory for systems where the exchange integral decreases rather slowly so that the spin exchange process may proceed effectively in a comparatively large region of interaction. If λ is the jump length of molecules in one diffusion step and $1/\varkappa$ is the distance at which the exchange interaction diminishes by a factor of e then, in the case

$$\lambda < 1/\varkappa \tag{2.225}$$

the approximation of sudden collisions is inapplicable and one needs an alternative model of the diffusive passage of the region of interaction by particles.

In the diffusive passage approximation the rate constant was calculated for the simplest system of particles with the spin $S = 1/2$ [2.67].

The Model

Just as before, the distance of closest approach of partners is assumed to equal the sum of their Van der Waals radii. The exchange integral is considered to be only dependent on a distance r between the colliding partners and independent of their mutual orientation. The decrease in the integral with the growth of r is approximated by the exponential function (2.45) (Fig.2.9). Just as in the model of sudden collisions, it is assumed that when two particles collide their spin interaction is sufficiently weak (less than kT) and does not significantly influence the kinematics of the collision. The colliding particles are assumed to enter and leave the region of interaction exclusively due to random walk performed independent of what is happening to the spins. In the process of diffusion the distance r between the partners changes randomly with time. As a result a random modulation of the exchange integral occurs, $J[r(t)] = J(t)$.

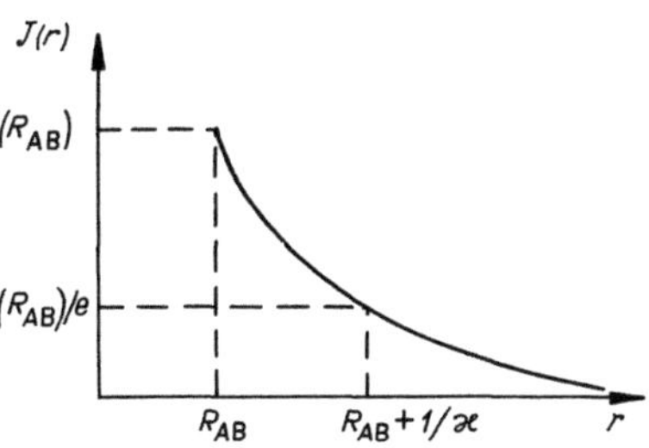

Fig. 2.9. Radial dependence of the exchange integral. R_{AB} is the sum of the Van der Waals radii of A and B particles, $1/\varkappa$ is the distance at which the exchange integral diminishes in e times

Thus the calculation of the exchange rate constant in the diffusion model is similar to that encountered in the previous section dealing with the random modulation of the partners' exchange interaction by their rotational motion.

An Equation for the Operator of Collision Efficiency

The operator $\hat{T}$ of collision efficiency satisfies (2.159) in which $\hat{\Omega}(\tau)$ is randomly modulated by the partners' diffusion. By following the reasoning that helped to pass from (2.159) to (2.161) we introduce a partial tensor $\hat{T}(\tau,r)$ of such pairs of colliding partners which at a moment τ are characterized by an assigned distance r between them. A full change of $\hat{T}(\tau,r)$ is assigned by the additive sum of contributions determined by the dynamics of spins at a given value of r and by the diffusion of particles resulting in a change in r. Similar to (2.161) we obtain [2.23]

$$(\partial/\partial\tau)\hat{T}(\tau,r) = -[\hat{T}(\tau,r), \hat{Q}] + \hat{\Omega}(r)\hat{T}(\tau,r) + D\Delta\hat{T}(\tau,r) , \qquad (2.226)$$

where Δ is Laplacian. Under the steady-state conditions the partial tensor satisfies the equations [see also (2.177)]

$$-[\hat{T}(r), \hat{Q}] + \hat{\Omega}(r)\hat{T}(r) + D\Delta\hat{T}(r) = 0 \quad . \qquad (2.227)$$

To formulate the boundary conditions for (2.227) we introduce a partial pair density matrix $\rho^{AB}(r,t)$ of a subensemble of colliding partners with an assigned distance r between them

$$\rho^{AB}(r,t) = \hat{T}(r)\rho^{A} \times \rho^{B} \quad . \qquad (2.228)$$

The boundary condition on infinity is given by [2.82]

$$\rho^{AB}(r,t) \to \rho^{A}(t) \times \rho^{B}(t) , \qquad r > r_i \quad . \qquad (2.229)$$

From (2.229) follows the condition

$$\hat{T}(r) \to \hat{E} , \qquad r > r_i \quad , \qquad (2.230)$$

where $\hat{E}$ is a unit operator. Zero diffusion flow through a sphere with a radius R_{AB} gives another boundary condition

$$D\nabla\hat{T}(r)|_{r=R_{AB}} = 0 \quad , \qquad (2.231)$$

where D is the partners' mutual diffusion coefficient.

The Kinetic Equations

The variation of one-particle density matrices ρ^A and ρ^B induced by pair collisions of A and B can be presented in the form of (2.177). For example, the variation of ρ^A due to spin exchange is given by the relation [2.97]

$$(\partial\rho^A/\partial t)_e = C_B \mathrm{Tr}_B\{ \int d^3\underline{r}\hat{\Omega}(r)\hat{T}(r)\rho^A \times \rho^B\} \quad . \tag{2.232}$$

In a number of cases the right-hand side of (2.232) can be presented as a flux of a two-particle matrix from the region of interaction. The commutator $[\hat{T}, \hat{Q}]$ of (2.227) makes a negligible contribution, of the order of $\delta\tau_c$ [cf with (2.136-140)], to the efficiency of exchange. Assuming $[\hat{T}, \hat{Q}] = 0$, substituting $\hat{\Omega}\hat{T}$ from (2.227) into (2.232), and passing from the volume integral to the surface one, we obtain [2.98]

$$(\partial\rho^A/\partial t)_e = -C_B D \mathrm{Tr}_B\{ \int_{S(r_i)} d\underline{s}\nabla\hat{T}(r)\rho^A \times \rho^B\} \ ,$$

$$(\partial\rho^B/\partial t)_e = -C_A D \mathrm{Tr}_A\{ \int_{S(r_i)} d\underline{s}\nabla\hat{T}(r)\rho^A \times \rho^B\} \ , \tag{2.233}$$

where $S(r_i)$ is the surface of the interaction sphere with a radius r_i. The flux of the two-particle matrix through $S(r_i)$ is no longer dependent on r_i if the latter is assumed to be sufficiently great. Therefore in (2.233) one can put $r_i \to \infty$.

The Spin Exchange Between Two Kinds of Particles with Spins $S_A = S_B = 1/2$

A spin-Hamiltonian of two isolated particles is chosen in the form

$$\hat{H}_0 = \hbar\omega_A\hat{S}_{Az} + \hbar\omega_B\hat{S}_{Bz} \tag{2.234}$$

and their exchange interaction is given by the spin Hamiltonian

$$\hat{V} = \hbar J(R_{AB}) \exp[-\varkappa(r - R_{AB})]\hat{S}_A\hat{S}_B \quad . \tag{2.235}$$

The eigenvectors of $\hat{H}_0$, $|k\rangle$, are an external product of the eigenfunctions of S_{Az} and S_{Bz}, $|m_A\rangle$ and $|m_B\rangle$, i.e., $|k\rangle = |m_A\rangle \times |m_B\rangle \equiv |m_A, m_B\rangle$. Denote the states of $\hat{H}_0$ as

$$|1\rangle = |\tfrac{1}{2},\tfrac{1}{2}\rangle \ ; \quad |2\rangle = |\tfrac{1}{2},-\tfrac{1}{2}\rangle \ ;$$

$$|3\rangle = |-\tfrac{1}{2},\tfrac{1}{2}\rangle \ ; \quad |4\rangle = |-\tfrac{1}{2},-\tfrac{1}{2}\rangle \quad . \tag{2.236}$$

The eigenstates for $\hat{V}$ are the singlet state $|S\rangle = (|2\rangle - |3\rangle)/\sqrt{2}$ and the states of the triplet $|T_+\rangle = |1\rangle$, $|T_-\rangle = |4\rangle$ and $|T_0\rangle = (|2\rangle + |3\rangle)/\sqrt{2}$. In the S-T basis $\hat{H}_0$ has the only nonzero nondiagonal matrix element that mixes up the states $|S\rangle$ and $|T_0\rangle$

$$(H_0)_{ST_0} = \hbar(\omega_A - \omega_B)/2 \equiv \hbar\delta/2 \quad . \tag{2.237}$$

If a singlet-triplet intersystem crossing due to the difference of Larmor frequencies of spins δ fails to reveal itself within the time of the diffusive passage of the interaction region by particles, i.e., if the condition

$$\delta r_i^2/D < 1 \tag{2.238}$$

is fulfilled, then in (2.227) one may neglect δ. According to [2.67] the influence of δ on spin exchange is negligible in the case

$$\delta R_{AB}^2/D < 1 \; ; \quad \delta/(D\varkappa^2) < 1 \; , \tag{2.239}$$

where $\varkappa$ characterizes an exchange integral decay (see Fig.2.9). Neglecting the effect of δ in the spin exchange process implies that the exchange is considered in an adiabatic approximation: in the process of collision the populations of the singlet and triplet states are preserved.

In the approximation $\delta = 0$ from (2.233) we find that the change of nondiagonal density matrix elements of A and B due to exchange is described by

$$(\partial/\partial t)\rho^A_{\frac{1}{2},-\frac{1}{2}} = -4\pi C_B D \int_{R_{AB}}^{\infty} dr \cdot r^2(\Delta T_{13,k\ell} + \Delta T_{24,k\ell}) \cdot (\rho^A \times \rho^B)_{k\ell} \; ,$$

$$(\partial/\partial t)\rho^B_{\frac{1}{2},-\frac{1}{2}} = -4\pi C_A D \int_{R_{AB}}^{\infty} dr \cdot r^2(\Delta T_{12,k\ell} + \Delta T_{34,k\ell}) \cdot (\rho^A \times \rho^B)_{k\ell} \quad . \tag{2.240}$$

The matrix elements $T_{mn,k\ell}$ determining the exchange efficiency satisfy a system of coupled equations of the kind

$$D\Delta T_{13,13} = (i/2)J(r) \cdot (T_{13,13} - T_{12,13}) \; ,$$
$$D\Delta T_{12,13} = (i/2)J(r) \cdot (T_{12,13} - T_{13,13}) \quad . \tag{2.241}$$

It is convenient to solve (2.241) in the S-T basis: we obtain a system of uncoupled equations

$$D\Delta T_{pq,rs} = i\hbar^{-1}[V_{pp}(r) - V_{qq}(r)]T_{pq,rs} \; , \tag{2.242}$$

where $|p\rangle$, $|q\rangle$, $|r\rangle$, $|s\rangle = |S\rangle$, $|T_+\rangle$, $|T_0\rangle$, $|T_-\rangle$. Equation (2.242) under the boundary conditions (2.230) and (2.231) yields nonzero solutions only for the diagonal elements $T_{pq,pq}$. Linearizing the right-hand sides of (2.240) with the help of $\rho = \rho_0 + \delta$ and assuming $\rho_{0_{nn}} \cong 1/2$ we arrive at

$$[(\partial/\partial t)\sigma^A_{\frac{1}{2},-\frac{1}{2}}]_e = -K_e C_B(\sigma^A_{\frac{1}{2},-\frac{1}{2}} - \sigma^B_{\frac{1}{2},-\frac{1}{2}}) ,$$

$$[(\partial/\partial t)\sigma^B_{\frac{1}{2},-\frac{1}{2}}]_e = -K_e C_A(\sigma^B_{\frac{1}{2}-\frac{1}{2}} - \sigma^A_{\frac{1}{2},-\frac{1}{2}}) \tag{2.243}$$

with the exchange rate constant

$$K_e = 4\pi r_e D , \tag{2.244}$$

where we introduce the effective exchange radius

$$r_e = (1/2)\mathrm{Re}\{\int_{R_{AB}}^{\infty} dr \quad r^2 \Delta T_{T_+S,T_+S}(r)\} . \tag{2.245}$$

The value of T_{T_+S,T_+S} is found from

$$D\Delta T_{T_+S,T_+S} - iJ(r)T_{T_+S,T_+S} = 0 .$$

Hence we obtain

$$T_{T_+S,T_+S} = (c_1/r)I_0(z) + (c_2/r)N_0(z) ,$$

where

$$z = z_0 \cdot \exp[-\varkappa(r - R_{AB})/2] , \quad z_0 = 2[-iJ(R_{AB})/(D\varkappa^2)]^{\frac{1}{2}}$$

while I_k and N_k are the Bessel functions of the first and second kind.

From the boundary conditions we find

$$c_1 = (\pi/\varkappa)[N_1(z_0)z_0\varkappa R_{AB} - 2N_0(z_0)]/[I_1(z_0)z_0\varkappa R_{AB} - 2I_0(z_0)] ,$$

$$c_2 = -\pi/\varkappa .$$

Substituting the solution obtained for T_{T_+S,T_+S} into (2.245) we obtain

$$r_e = R_{AB}/2 + \mathrm{Re}\{\{[C + \ln(z_0/2)]/\varkappa - c_1/2\}\} , \tag{2.246}$$

where C is the Euler constant.

Figure 2.10 lists the values of r_e in limiting situations. The abscissa characterizes the rate of the drop of the exchange integral while the ordinate

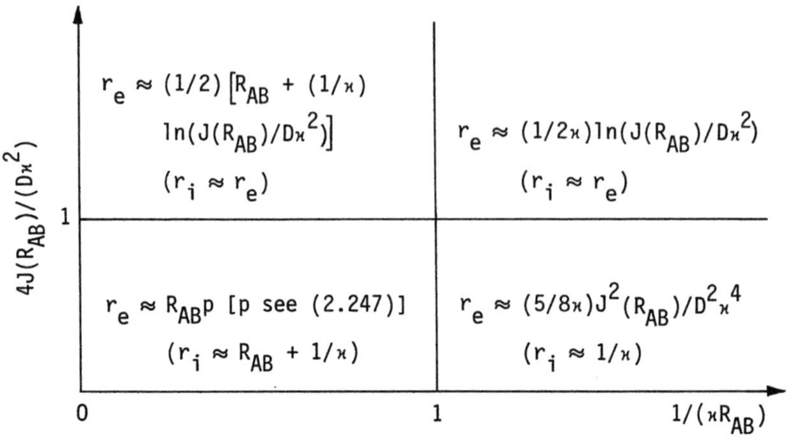

Fig. 2.10. Values of the effective radius of exchange in the model of diffusive passage of the interaction region. The formulae are given for the extreme values of parameters, characterizing the magnitude of exchange interaction (ordinate) and the steepness of the exchange integral decrease with the distance between the particles (abscissa)

assigns the quantity $|z_0|^2$. When the exchange integral drops sharply, i.e., $\varkappa > 1/R_{AB}$, then in the case of a relatively weak interaction of particles, when $|z_0|^2 = 4J(R_{AB})/D\varkappa^2 < 1$, the spin exchange occurs practically only at a close approach of the particles in a narrow layer between spheres with the radii R_{AB} and $R_{AB} + 1/\varkappa$. In this case in the diffusive model one can introduce an exchange efficiency at one collision as was the case in the model of sudden collisions. Indeed, when the conditions $\varkappa R_{AB} > 1$ and $|z_0|^2 < 1$ are fulfilled r_e can be presented as the product of R_{AB} and of the exchange efficiency p

$$r_e = R_{AB} \cdot p \quad ,$$

$$p = (1/2)J^2(R_{AB}) \cdot \bar{\tau}^2/[1 + J^2(R_{AB}) \cdot \bar{\tau}^2] \quad , \tag{2.247}$$

where the effective time of interaction is

$$\bar{\tau} = R_{AB}/(D\varkappa) \quad . \tag{2.248}$$

According to (2.247) $r_e \to R_{AB}/2$ with the growth of the exchange integral $J(R_{AB})$ at a minimum approach of the particles. However a further increase in the spin interaction, when $|z_0|^2 > 1$, ensures a spin exchange not only at the minimum distance of approach of the particles, but also at distances $r > R_{AB}$. Therefore the effective radius at which an intensive exchange already occurs becomes greater than R_{AB}. In this case the spin exchange rate constant is given by the relation

$$K_e = 4\pi D\{R_{AB}/2 + (1/2\varkappa) \cdot \ln[|J(R_{AB})|/D\varkappa^2]\} \quad . \tag{2.249}$$

Note should be taken of the logarithmic dependence of the effective radius of exchange on the coefficient of diffusion.

In the reverse situation of the slow decline of the exchange integral, when $\varkappa < 1/R_{AB}$, the spin exchange represents an integral effect of interaction as the particles pass a relatively large region with a radius $r_i \sim 1/\varkappa$. The result is that, as expected, r_e practically no longer depends on the distance of the minimum approach of the particles (see Fig.2.10).

Comparison of Results Obtained from Models of Sudden Collisions and Those of Diffusive Passage

In the model of sudden collisions the effective radius of exchange for particles with the spin S = 1/2 is equal to [see (2.144)]

$$r_e = (R_{AB}/2)J^2(R_{AB}) \cdot \tau_c^2/[1 + J^2(R_{AB}) \cdot \tau_c^2] \quad . \tag{2.250}$$

Comparison of (2.250) with the results of the diffusive model presented in Fig.2.10 shows that it is only at a sharp drop of the exchange integral with the growth of distance between spins and at moderate values of $J(R_{AB})$ when $|z_0|^2 < 1$ that there is full agreement between the results of the two models. In both cases the effective radius of exchange may be represented as the product of the distance of the closest approach of particles by the exchange efficiency for one collision, $r_e = R_{AB} \cdot p$ [cf (2.247) with (2.250)]. In other cases (see Fig. 2.10) the spin exchange is essentially contributed to not only by the interaction at the minimum approach of particles but also by that at longer distances between them. Then the concept of the collision efficiency p becomes invalid; the latter may be even larger than 1, if it is formally introduced by the relation $p = r_e/R_{AB}$.

The main difference between the models in question is that the sudden collision model takes into account only the exchange at the minimum distances between particles while the diffusive one considers that at various distances. The spin exchange at various distances between the particles can be taken into consideration in the model of sudden collisions too, if it is appropriately generalized. Let the molecules diffuse in a solution by long jumps so that they get into the region of interaction or leave it in one jump. But we shall assume that at the moment the exchange interaction is switched on the spins may find themselves not only at the distance R_{AB} but also at longer ones within the range $R_{AB} \leq r \leq r_i$. The exchange efficiency in spin collisions at the distance r is $p(r) = (1/2)J^2(r) \cdot \tau_c^2/(1 + J^2(r) \cdot \tau_c^2)$, and the probability that in the region of interaction the partners will be in the layer between two spheres with the radii r and $r + dr$ is given by the equation

$$dw(r) = 4\pi r^2 dr/v(R_{AB}, r_i) \quad .$$

The effective exchange radius averaged over all possible collisions is equal to

$$r_e = \int_{R_{AB}}^{r_i} rp(r)dw(r) = (1/2) \int_{R_{AB}}^{r_i} \{rJ^2(r)\tau_c^2/[1 + J^2(r)\tau_c^2]\}dw(r) \quad . \tag{2.251}$$

In the case of weak exchange when $J(R_{AB})\tau_c < 1$ and the effective exchange occurs in the layer between the spheres with the radii R_{AB} and $R_{AB} + 1/\varkappa$ from (2.251) we obtain the following results [2.67]:

$$r_e \cong (1/2)R_{AB}J^2(R_{AB}) \cdot \tau_c^2 \quad , \qquad \text{if } \varkappa R_{AB} > 1$$

and

$$r_e \cong (1/2\varkappa)J^2(R_{AB}) \cdot \tau_c^2 \quad , \qquad \text{if } \varkappa R_{AB} < 1 \quad . \tag{2.252}$$

When the exchange becomes strong, i.e., $J(R_{AB})\tau_c > 1$, then, taking into account the exponentially fast drop of $J(r)$, in the integrand of (2.251) one can assume the following approximation:

$$J^2(r)\tau_c^2/(1 + J^2(r) \cdot \tau_c^2) \cong \begin{cases} 1 \, , & \text{if } r \leq r_i \\ 0 \, , & \text{if } r > r_i \end{cases} \quad , \tag{2.253}$$

where r_i is determined from the condition

$$J(r_i) \cdot \tau_c = 1 \quad . \tag{2.254}$$

Hence we obtain

$$r_i = R_{AB} + (1/\varkappa) \ln[|J(R_{AB})| \cdot \tau_c] \quad . \tag{2.255}$$

Thus in the range of strong interaction we obtain

$$r_e = (1/2)\{R_{AB} + (1/\varkappa) \ln[|J(R_{AB})| \cdot \tau_c]\} \quad . \tag{2.256}$$

The appearance of the logarithmic dependence of r_e on $J(R_{AB}) \cdot \tau_c$ is due to the choice of the exponential law for the drop in $J(r)$. The comparison of

(2.252, 256) with the data of the diffusive model shows that the suggested generalization of the model of sudden collisions makes it possible to establish accord between the two models. Both models lead to the same dependences of the effective radius of exchange on the coefficient of diffusion and on the magnitude of the exchange integral at the closest approach of the particles. However, the characteristic time of molecule interaction for these models is naturally different: in the model of sudden collisions this is the time τ_c of the "settled" life of molecules (2.118) while in the diffusive model it is the time of diffusive passage of the interaction region by the particles. Thus, within the framework of the model of sudden switch-on one can obtain qualitatively correct results for the rate of spin exchange even if the decrease in the exchange integral with the increase in the distance between the partners takes place rather slowly and the exchange is efficiently performed over a comparatively large region.

An important task of research into the process of spin exchange in solutions is to obtain data on the overlap of the electron orbitals of colliding partners. It is noteworthy that the diffusive model and the generalized model of sudden switchings, in comparison with the model of sudden collisions at the closest approach of spins (Sect.2.5), yield qualitatively different answers to the question as to the possibilities of the spin exchange method for solving this problem. According to the model of sudden collisions, as the overlap of orbitals increases, in the case of strong exchange the effective radius of exchange tends to a certain finite value and ceases to be dependent on the scope and duration of interaction; there arises a sort of saturation with spin exchange. And only in the case of weak exchange does the dependence of the efficiency of exchange on the overlap of orbitals persist. But the results obtained in (Sect.2.6) show that in fact the analysis of the spin exchange process may yield the necessary evidence on the overlap of electron orbitals both in the case of weak, and that of strong, exchange.

At present we do not know much about the mangitude of the exchange integral, its radial and angular dependence for those real solutions of paramagnetic particles in which the process of spin exchange is studied. Therefore the above theory of spin exchange in solutions is based on the model potentials of interaction. One can try to estimate the parameters of these potentials by using the results of theoretical calculations for the simplest systems and by means of semiempirical methods (see Sect.2.1). However, the main source of information about the exchange interaction of paramagnetic particles colliding in solution is the experimental data on spin exchange. Analysis of experimental results with the help of model theories allows one to establish the main characteristics of the exchange interaction of particles in solutions. Therefore the numerical estimates of the exchange rate constants for various systems are reported not here but in Chap.4 where the experimental data are discussed.

3. Experimental Measurement of Spin Exchange Rate

The rate of spin exchange can be measured from any of its spectroscopic manifestations discussed in the previous chapter. Among the various methods which can be used for such measurements the ones most widely used are based on the analysis of the ESR line shape. These procedures are relatively simple and most accessible since they can be done with the help of routine ESR spectrometers. So these procedures will be considered here in more detail. The techniques based on other spectroscopic manifestations (saturation, electron-electron double resonance, spin echo) which have not yet been so widely used will be considered in less detail. Analysis of spectroscopic manifestations of spin exchange is often complicated by the interference of dipole-dipole interactions. Therefore the exclusion of dipole-dipole interactions or estimation of their contribution is an important problem and it will also be discussed in detail. Finally, methods will be considered of distinguishing between the cases of strong and weak exchange which is of great importance for the interpretation of data on spin exchange.

3.1 Experimental Measurement of Spin Exchange Rate Constants from ESR Spectra

As has been shown in the previous chapter, spin exchange may produce broadening, shift, or narrowing of the ESR lines. Any of these effects may be used to determine spin exchange rate constants. Note that the details of the analysis are somewhat different depending on whether the particles participating in spin exchange are identical or different.

3.1.1 Exchange Between Identical Particles with $S = 1/2$

Doublet Spectrum with a Zero Initial Width

Consider first the simplest case of spin exchange between identical paramagnetic particles with the electron spin $S = 1/2$ and the ESR spectrum consisting of two infinitely narrow lines of the same intensity (Fig.3.1a). These lines may, for example, be components of the hyperfine structure from a nucleus with a spin $I = 1/2$. For the spectrum in question the statistical weight of each component $\varphi = 1/2$.

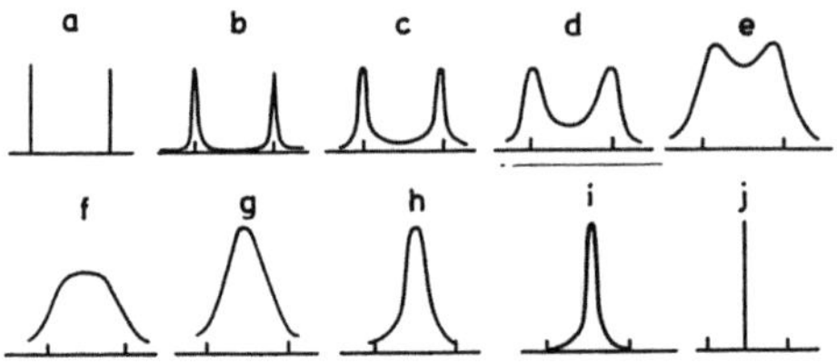

Fig.3.1a-j. Changes in the profile of a spectrum consisting of two infinitely narrow lines of the same intensity as the spin exchange rate increases (intensities are depicted in different scales)

In this case from (2.77) one can obtain the following general expression for the spectrum profile as a function of the exchange frequency $\omega_e = K_e \cdot C$

$$g(\omega) = k \cdot (a^2/K_e \cdot C) \cdot [\omega^2 + (\omega^2 - a^2/4)^2/K_e^2 \cdot C^2]^{-1} \quad . \qquad (3.1)$$

Here a (rad/s) is the hfi constant; ω (rad/s) is the current frequency counted from the spectrum center; K_e (l/mol·s) is the spin exchange rate constant; C (mol/l) is the concentration of paramagnetic particles; and k is the normalizing constant.

From (3.1), in accordance with (2.79, 89) it follows that in the region of slow exchange when

$$K_e \cdot C < a \qquad (3.2)$$

the spectrum consists of two identical lines with a shape close to a Lorentzian one. The halfwidth of these lines at a half-height $\Delta\omega_{1/2}$ (rad/s) is given by the formula[1]

$$\Delta\omega_{1/2} = K_e \cdot C \cdot (1 - \varphi) = K_e C/2 \qquad (3.3)$$

and their absorption maxima are shifted from their initial positions towards the spectrum center by the magnitude $\delta\omega$ (rad/s)

$$\delta\omega = (K_e C)^2 \cdot \varphi/a = (K_e C)^2/2a \quad . \qquad (3.4)$$

---

1 It should be emphasized that while the left-hand term of (3.3) contains a quantity $\Delta\omega_{1/2}$ with the dimension of angular frequency (rad/s), the right-hand term contains a quantity $K_e C \cdot (1 - \varphi)$ that characterizes the frequency of collisions effective in spin exchange and has a dimension of s^{-1}.

Note that the factors $1-\varphi$ and φ take into account the fact that spin exchange between particles with identical projections of the nuclear spin does not produce any change in the ESR spectrum.

As the frequency of spin exchange increases, the lines increasingly overlap (Fig.3.1b-e) and in the intermediate region of exchange frequencies

$$K_eC \approx a \tag{3.5}$$

they merge into one line (Fig.3.1f). In the region of fast exchange,

$$K_eC > a \quad , \tag{3.6}$$

the spectrum as seen from (3.1) represents a single Lorentzian line with a maximum at $\omega = 0$. The halfwidth of this line in conformity with (2.96) decreases with the growth of exchange frequency according to the equation

$$\Delta\omega_{1/2} = a^2/4K_eC \quad . \tag{3.7}$$

Thus, slow spin exchange produces broadening and shift of the hf components while fast exchange produces narrowing of the spectrum. The spin exchange rate constant can be determined from the analysis of any of the three effects.

The rate constant K_e can be calculated from expcrimentally measured characteristics of the doublet spectrum with a zero initial width with the help of the following formulas

1) from the line width in the region of slow exchange

$$K_e = \Delta\omega_{1/2}/(1-\varphi)\cdot C,\ \varphi = 1/2 \quad , \tag{3.8}$$

2) from the line shift in the region of slow exchange

$$K_e = (a\cdot\delta\omega)^{1/2}/\varphi^{1/2}\cdot C,\ \varphi = 1/2 \quad , \tag{3.9}$$

3) from the line width in the region of fast exchange

$$K_e = a^2/4\Delta\omega_{1/2}\cdot C \quad . \tag{3.10}$$

In practice the ESR line width is usually measured between the points of the maximum slope in units of magnetic induction (Gauss). Denote this quantity measured in Gauss as ΔH. It is convenient to measure the hyperfine splitting and exchange shift also in Gauss. Denote these parameters by ΔH_{hf} and δH, respectively.

It should be borne in mind that for lines of Lorentzian shape the width between the points of the maximum slope ΔH is connected with the halfwidth at the half-height $\Delta H_{\frac{1}{2}}$ by the relation

$$\Delta H = 2\Delta H_{\frac{1}{2}}/3^{\frac{1}{2}} \tag{3.11}$$

and the conversion coefficient for transforming the halfwidth in rad/s to that in Gauss is

$$\Delta\omega_{\frac{1}{2}}/\Delta H_{\frac{1}{2}} = g\beta/\hbar = 0.879 \cdot 10^7 \cdot g \quad , \tag{3.12}$$

where g is the g value of a paramagnetic particle; $\beta = 0.9273 \cdot 10^{-20}$ erg/G is the Bohr magneton; and $h = 6.625 \cdot 10^{-27}$ erg·s is Planck's constant; $\hbar = h/2\pi$. Hence one can rewrite expressions (3.8 - 10) for K_e as follows:

1) exchange broadening

$$K_e = 1.52 \cdot 10^7 (g/2(1-\varphi)) \cdot \Delta H/C, \ \varphi = 1/2 \quad , \tag{3.13}$$

2) exchange shift

$$K_e = 1.76 \cdot 10^7 (g/2\sqrt{\varphi}) \cdot (\Delta H_{hf} \cdot \delta H)^{\frac{1}{2}}/C, \ \varphi = 1/2 \quad , \tag{3.14}$$

3) exchange narrowing

$$K_e = 0.508 \cdot 10^7 (g/2)\Delta H_{hf}^2/\Delta H \cdot C \quad . \tag{3.15}$$

Substituting into (3.13 - 15) the concentration expressed in mol/l, and ΔH, ΔH_{hf}, and δH in Gauss one obtains K_e in l/mol·s.

Complex Spectra

In practice one usually has to calculate the spin exchange rate constants from spectra more complicated than the one considered above. The complications may be, first, due to the multicomponent hyperfine structure and, secondly, due to a nonzero initial line width in the absence of exchange. In these more complicated cases the spectrum profile $g(\omega)$ can still be found from (2.77), but it will be described by more cumbersome formulas than (3.1). Nevertheless, as has been shown in Sect.2.4, the influence of spin exchange on the spectra in the limiting situations of slow and fast exchange is qualitatively the same as for the doublet spectrum. In the case of slow exchange a broadening of individual hyperfine components occurs while in the case of fast exchange one observes a single line of Lorentzian shape which becomes narrower as the exchange rate increases (Fig.3.2).

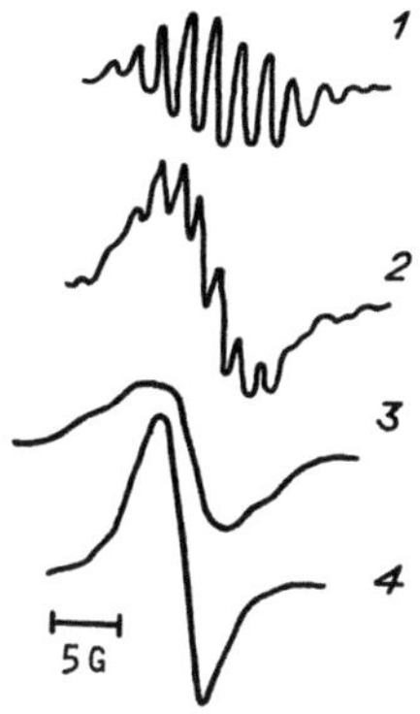

Fig.3.2. Change in the ESR spectrum profile with the increase of spin exchange frequency for biphenyl anionradical in diethylene glycol dimethyl ether [3.1]

As a practical criterion of slow exchange for a spectrum with a complex hyperfine structure the following condition may be used:

$$K_eC < a_{min} \quad , \tag{3.16}$$

where a_{min} is the minimum hyperfine coupling constant of the spectrum. As a criterion of fast exchange one may use the condition

$$K_eC > a_{max} \quad , \tag{3.17}$$

where a_{max} is the maximum hyperfine coupling constant of the spectrum. Thus, the complication of hyperfine structure results in the broadening of the region of intermediate exchange

$$a_{min} < K_eC < a_{max} \quad . \tag{3.18}$$

In this region the calculation of rate constants turns out to be practically feasible only by using computers.

The presence in the initial spectrum of a nonzero line width of nonexchange origin must be taken into account in measuring K_e from both the line width and the shift. The procedures for determining spin exchange rate constants from complex spectra in various practically important situations will be considered below.

Determination of Rate Constants from Exchange Broadening of Complex Spectra

As has been pointed out earlier, the criterion of slow exchange for a spectrum with a complex hyperfine structure is given by (3.16). If in the absence of exchange the spectral lines are of Lorentzian shape then the exchange-broadened lines in the region of slow exchange in the first approximation remain Lorentzian and their width between the points of the maximum slope can be presented as

$$\Delta H^{(k)} = \Delta H_0^{(k)} + \Delta H_e^{(k)} \quad . \tag{3.19}$$

Here k is the index of the line; ΔH_0, the initial width in the absence of exchange; ΔH_e, the line broadening due to spin exchange (exchange broadening). Note that both the initial width ΔH_0 and the exchange broadening may be different for different hyperfine components. We shall always imply this in further discussion though for brevity in formulas we shall omit the index k.

Just as in the case of the doublet spectrum, the spin exchange rate constant may be calculated from exchange broadening with the help of equations similar to (3.13)

$$K_e = 1.52 \cdot 10^7 [g/2(1 - \varphi)] \Delta H_e / C \tag{3.20}$$

or

$$K_e = 1.52 \cdot 10^7 [g/2(1 - \varphi)] (\Delta H - \Delta H_0) / C \quad . \tag{3.21}$$

The statistical weights $\varphi(k)$ can be found from the relations

$$\varphi(k) = S^{(k)}/S \quad , \tag{3.22}$$

where $S^{(k)}$ is the theoretical intensity of the component from whose broadening K_e is calculated and S is the sum of theoretical intensities of all the components of the spectrum.

Note that in some papers the spin exchange rate constant was calculated from exchange broadening by using the relation

$$K_b = 1.52 \cdot 10^7 (\Delta H - \Delta H_0) / C \tag{3.23}$$

rather than (3.21), i.e., one obtained, in fact, the broadening constant $K_b = [2(1 - \varphi)/g] K_e$ rather than the exchange rate constant K_e. Because of a possible difference between the values of φ for the different hyperfine components the parameter K_b for one and the same process may also depend on the line of the spectrum for which ΔH is measured. Hence, in contrast to K_e, the constant K_b characterizes not only the rate of spin exchange but also the way of measuring this value.

For the unity of notation in this monograph all the constants K_b calculated in the original articles with the help of (3.23) have been recalculated into K_e. Because of this for some processes the values of exchange rate constant that we cite differ from those cited in the literature.

In practice K_e is usually determined from the measured dependence of the line width on the concentration of paramagnetic particles. From (3.21) it follows that the width ΔH is a linear function of the concentration of paramagnetic particles. Therefore K_e can be calculated from the slope of the dependence of ΔH on C. As an example Fig. 3.3 presents such a dependence for bis-acetylacetonate of vanadyl VO $(acac)_2$ in a pyridine solution. For this compound $g = 1.97$, the factor $1 - \varphi = 7/8$ is the same for all hyperfine components, and the initial hyperfine lines in diluted solutions have a shape close to the Lorentzian one. The rate constant K_e of spin exchange between $VO(acac)_2$ particles, as calculated, with the help of (3.21), from the data presented in this figure, is equal to $(8.0 \pm 0.3) \cdot 10^8$ 1/mol·s.

Numerous experimental data indicate that the dependence of the total line width ΔH on the concentration of paramagnetic particles is usually well described by (3.21) until ΔH reaches 30-40% of the distance between the lines. For example, for $VO(acac)_2$, as seen from Fig.3.3 ΔH amounts to ~30% of ΔH_{hf}, but no deviation from linearity is observed yet. Simulation of exchange-broadened spectra with the help of computers gives additional support to this conclusion. As an example Fig.3.4 presents the results of computer calculations of the line width for the central component of the spectrum consisting of three hyperfine lines of the same intensity as a function of the product $K_e \cdot C$. Such spectra are characteristic, e.g., of nitroxyl free radicals that are often used in experimental studies of spin exchange. The values of a hyperfine coupling constant expressed in (G) and (rad/s) are used in Fig.3.4

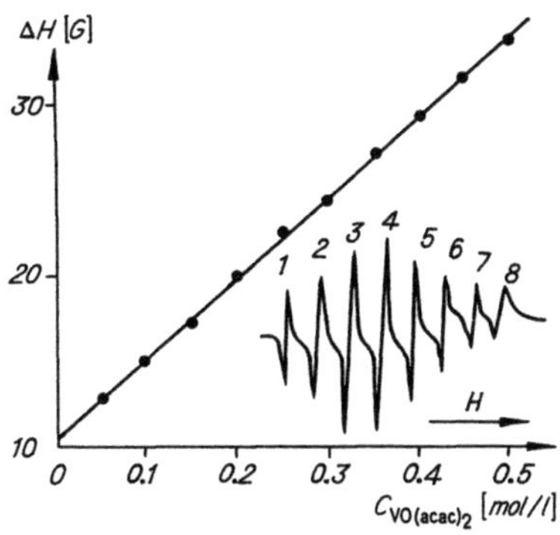

Fig.3.3. The ESR spectrum of $VO(acac)_2$ in a pyridine solution and the concentration broadening of its 4th hyperfine component [3.2]

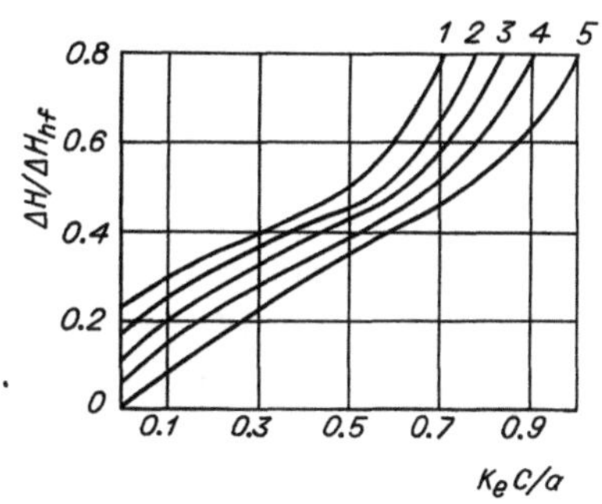

Fig.3.4. Dependence of the width of the central hyperfine component of a triplet spectrum on the parameter K_eC/a [3.3]. Values of $\Delta H_0/\Delta H_{hf}$: (1) 0.24, (2) 0.17; (3) 0.12; (4) 0.06; (5) 0.00

as unit measures of the line width $\Delta H(G)$ and of the frequency $\omega_e = K_e C$ (rad/s) which allows the results of these calculations to be used for the analysis of three-component spectra with arbitrary splittings. Various curves correspond to different initial widths, the lines at $\omega_e = 0$ being assumed to have a Lorentzian shape. The linear relationship between ΔH and C is indeed seen to be conserved with a sufficient accuracy up to $\Delta H \approx 0.3\Delta H_{hf}$ for any initial width. Note that the data depicted in Fig.3.4 may certainly be used for measuring exchange frequency from experimental spectra not only within the region of $\Delta H \lesssim 0.3\Delta H_{hf}$ but also over the entire range of ω_e and ΔH variations. In order to do so, however, it is necessary to determine the initial width ΔH_0 from independent experiments at low exchange frequencies.

Note that deviations from the linear dependence between the experimentally measured line width and the frequency ω_e at $\Delta H > 0.3\Delta H_{hf}$ are induced not only by an overlap of broadened components, but also by a deviation of the observed spectrum profile from the sum of the three Lorentzian components as a result of spin exchange. For this reason when analyzing the exchange-broadened spectrum at $\Delta H > 0.3\Delta H_{hf}$ one cannot use the Atlas of ESR spectra [3.4] which takes into account only the effect of hyperfine component overlapping.

In calculating the rate constants from exchange broadening certain difficulties may arise when the lines of the initial spectrum have a non-Lorentzian shape due to an unresolved hyperfine structure. In this case spin exchange brings about not only a broadening of experimentally observed lines but also a change in their shape. As a result, the dependence of width on concentration will be linear only at sufficiently large concentrations while at small concentrations marked deviations from linearity may be observed. The rate constant should naturally be determined from the slope of the linear part of the curve $\Delta H = f(C)$. The measurement of K_e from the initial nonlinear part usually requires the use of computers. An example of such an investigation is [3.5] where the rate constants of spin exchange between di-tert-butylnitroxide (DTBN) radicals in binary solutions were measured. Note that in this case the calculation of the line profile cannot be reduced to taking account of the influence of the finite width of the unresolved hyperfine components, say, by the technique described in [3.6]. For closely located unresolved hyperfine components even at low concentrations the intermediate or fast exchange conditions may be fulfilled and, hence, the spectrum profile may differ from a simple superposition of individual hyperfine lines. As an example Fig.3.5 represents the concentration dependence, obtained by NIKITAYEV, of the line width of a stable free radical of 2,2,6,6-tetramethyl-4-oxopiperidine-1-oxyl (TANONE) in an aqueous solution.

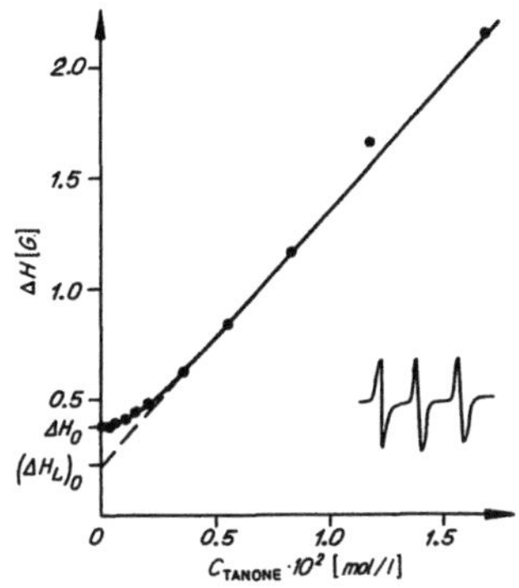

Fig.3.5. The ESR spectrum of TANONE in water solution and the concentration broadening of its central hyperfine component. Deviations from the linear dependence at small concentrations arise due to the non-Lorentzian shape of the initial line which originates from the unresolved hyperfine structure

This dependence is linear in the range of concentrations $(0.4 \div 1.7) \cdot 10^{-2}$ mol/l, but it has a nonlinear part at concentrations $<0.3 \cdot 10^{-2}$ mol/l. This is due to the fact that all the three hyperfine components from a nitrogen nucleus (hyperfine coupling constant $\Delta H_{hf}^{(N)} = 15.6$G) of the initial spectrum of TANONE contain an unresolved hyperfine structure from 12 protons of the methyl groups ($\Delta H_{hf}^{(H)} \approx 0.1$ G). At small concentrations (the initial nonlinear part of the curve) the spin exchange induces simultaneous broadening of the lines and averaging out of the unresolved proton hyperfine structure. The linear part corresponds to the situation in which the unresolved hyperfine structure has been already averaged out. The concentration dependence of ΔH corresponds to the exchange broadening of hyperfine components from the nitrogen nucleus with an initial Lorentzian width $(\Delta H_L)_0$ (see Fig.3.5). Taking into consideration that for TANONE $1 - \varphi = 2/3$ for all the three components and $g = 2.00$ one can obtain from the linear part of the dependence between ΔH and C, $K_e = 2.7 \cdot 10^9$ l/mol·s.

The described technique of measuring spin exchange rate constants is based on the assumption that spin exchange makes a predominant contribution to broadening of ESR lines as the concentration of paramagnetic particles increases. But line broadening can be caused by the dipole-dipole interaction of paramagnetic particles as well. The methods that enable one to establish whether the exchange contribution to broadening is indeed a predominant one will be discussed in Sect.3.3. Here it is only worth noting that for all the systems discussed in this section the dipole-dipole contribution to broadening is indeed negligible as compared to the spin exchange contribution.

Determination of Rate Constants from Exchange Shift of the Lines of Complex Spectra

For ESR spectra with a complex hyperfine structure the spin exchange rate constants may be calculated from an exchange shift of lines with the help of (2.87). Substituting in this formula the quantities expressed in angular frequency units by those expressed in units of magnetic induction one obtains

$$\delta H^{(k)} = [(\sqrt{3}/2)\Delta H_0^{(k)} + 1.14 \cdot 10^{-7} \cdot K_e C/g]$$

$$\times\ 1.14 \cdot 10^{-7}(K_e C/g) \sum_{n(\neq k)} \varphi(n)/\Delta H_{hf}^{(nk)} \quad . \qquad (3.24)$$

Here $\delta H^{(k)}$(G) is the shift of the kth hyperfine component; $\Delta H_0^{(k)}$(G) its initial width between the points of the maximum slope; $\Delta H_{hf}^{(nk)}$(G) the distance between the kth and the nth hyperfine components; $\varphi(k)$ and $\varphi(n)$ their statistical weights calculated according (3.22); and $K_e C$ is expressed in s^{-1}. The summation is performed over all components except the kth one. The initial lines are assumed to have a Lorentzian shape resulting from relaxation.

As an example consider the ESR spectrum of $VOSO_4$ (Fig.3.6) with $g = 1.96$ and $\Delta H_{hf} = 118$ G. The spectrum consists of 8 hyperfine components of the same intensity. Consequently, for every component $\varphi = 1/8$. Within the studied range of concentrations the shift which is quadratic in the concentration was found to prevail for $VOSO_4$.

The outer hyperfine components 1 and 8 are shifted most strongly; therefore it is reasonable to measure K_e from the change of the splitting between them

$$\delta H = 2\delta H^{(1)} = [\sqrt{3}\Delta H_0 + 2.28 \cdot 10^{-7} K_e C/1.96]$$

$$\times\ 1.14 \cdot 10^{-7} K_e C \sum_{n=1}^{7} (1/n)/(1.96 \cdot 8 \cdot 118) \quad . \qquad (3.25)$$

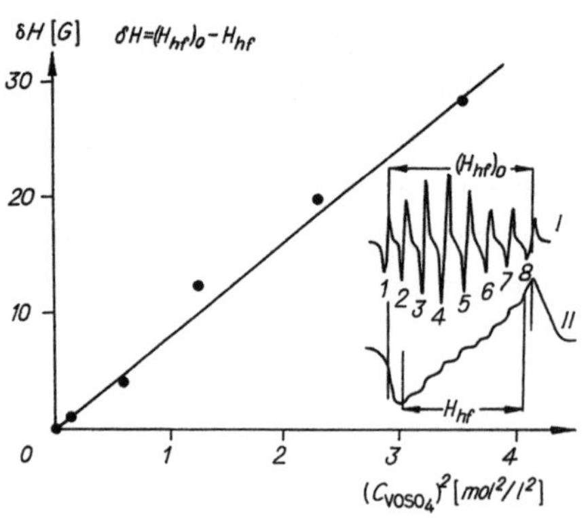

Fig.3.6. Concentration dependence of the exchange shift and the ESR spectrum of $VOSO_4$ in an aqueous solution [3.7]. (I) Initial spectrum in the absence of exchange; (II) spectrum with the maximum shift. The exchange shift is proportional to the square of the paramagnetic particles concentration

From the data presented in Fig.3.6 with the aid of (3.25), one obtains the constant $K_e = (8 \pm 4)\cdot 10^8$ l/mol·s.

It should be noted that from the dependence of line shift on concentration the spin exchange rate constants are generally determined with a smaller accuracy than from exchange broadening. The measurements of K_e from line shift, however, are of great importance for confirming the theory of spin exchange.

Determination of Rate Constants from Exchange Narrowing of Complex Spectra

To calculate K_e from the width of the exchange-narrowed line of a complex spectrum one should make use of (2.96). Taking into account that

$$< \Delta\omega^2 > = \sum_i I_i(I_i + 1)a_i^2/3 \qquad (3.26)$$

where a_i(rad/s) is the hyperfine coupling constant from the ith nucleus, I_i is the spin of this nucleus, and substituting into this formula the values measured in units of magnetic induction we obtain

$$\Delta H = \Delta H_0 + 3.39 \cdot 10^6 g \sum_i I_i(I_i + 1)(\Delta H_{hf}^{(i)})^2/K_e C \quad . \qquad (3.27)$$

It might seem that the rate constants K_e can be conveniently measured from the dependence of experimentally observed line width on reciprocal concentration which, according to (3.27), must be linear. However, the problem is that for many systems exchange narrowing can only be observed at relatively high concentrations of paramagnetic particles when the exchange-independent contribution to the line width ΔH_0 can no longer be considered a constant parameter. The width ΔH_0 can vary with concentration, e.g., due to the variation of the solution viscosity, to the association of paramagnetic particles in the solution, and to the influence of dipole-dipole interaction, the role of the latter increasing with the concentration of paramagnetic particles. Besides, in concentrated solutions the spin exchange rate constant may itself become dependent on concentration. As a result, the dependence of line width on concentration may become by far more complicated than reflected in (3.27) with constant values of ΔH_0 and K_e.

An example of such a dependence is given in Fig.3.7. Its nonmonotonous character appears to be due to the influence of dipole-dipole interactions on ΔH which becomes particularly important in the region of high concentrations ($C > 0.5$ mol/l). Therefore the rate constant of the spin exchange between DTBN radicals should be estimated from the slope of the dependence of ΔH on $1/C$ at lower concentrations ($C = 0.06 - 0.30$ mol/l) when the contribu-

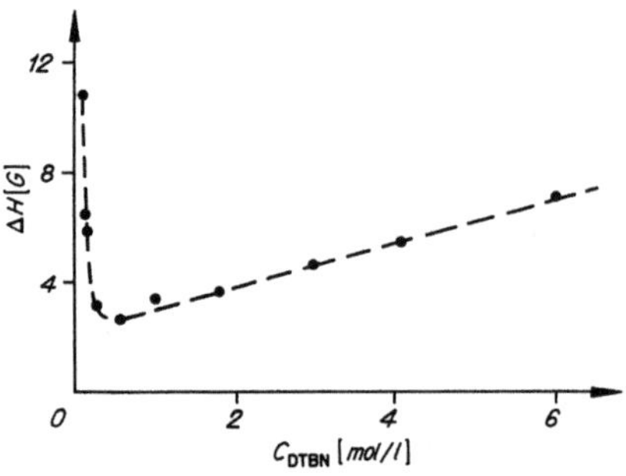

Fig.3.7. Concentration dependence of the line width in the region of exchange narrowing for di-t-butilnitroxide free radical in dimethylformamide [3.8]. The initial decrease of the line width is a result of the exchange narrowing. Its increase at still higher concentrations arises from the dipole-dipole interactions

tion of dipole-dipole interaction to ΔH is far less than that of spin exchange. Taking into account that for DTBN $g = 2.00$, the nuclear spin of nitrogen $I = 1$, $\Delta H_{hf} = 15.5$ G, from the data of Fig.3.7 one obtains the constant $K_e = (3.8 \pm 0.3) \cdot 10^9$ l/mol·s.

Peculiarities of Exchange Narrowing in Concentrated Solutions

When estimating spin exchange rate constants from ESR line narrowing in concentrated ($C \gtrsim 1$ mol/l) solutions one should be aware that the mechanism of exchange narrowing for such solutions may differ from that described in Sect. 2.4. Indeed, at $C \gtrsim 1$ mol/l the mean distance between paramagnetic particles becomes comparable with molecular diameters. As a result, these particles may for a significant part of time be in the state of collision, i.e., they may be located near one another forming stochastic (random) clusters – couples, triples, quadruples, etc. – for which

$$J^2 > \langle \Delta\omega^2 \rangle \quad , \tag{3.28}$$

where J is an exchange integral for interaction between the particles within a cluster and Δω is the difference of resonance frequencies that correspond to different lines of the ESR spectrum. In such solutions an exchange narrowing of ESR lines may occur via the same mechanism as in solids [3.9]. For this mechanism no translational motion of paramagnetic particles is necessary. That is why this mechanism of narrowing may be called a static one.

The possibility of such a mechanism of exchange narrowing in concentrated liquid solutions was first indicated in the theoretical papers [3.10, 11]. Experimentally this problem was studied in detail [3.7, 12-15] for solutions of various complexes of bivalent copper and vanadyl. It was found that in most cases, though not always, the dependence of the width ΔH of the exchange-narrowed ESR spectra of these compounds on the concentration C is described by the equation

$$\Delta H = \Delta H_0 + A/C \quad , \tag{3.29}$$

where the parameters ΔH_0 and A are independent of concentration. Figure 3.8 depicts such a dependence for spin exchange between $Cu(NH_3)_5^{2+}$ complexes in an aqueous solution containing 10 M NH_4OH. It would seem that, according to (3.27), such a dependence might be treated as a result of exchange narrowing by means of binary collisions of paramagnetic particles. However, it is not so since the width of an exchange-narrowed line, and hence the coefficient A, do not depend on the temperature and viscosity of the solution (Fig.3.9). Besides, for $Cu(NH_3)_5^{2+}$ and vanadyl salts the exchange narrowing does not disappear even when the solution is cooled below the temperature of vitrification. These data testify to the noncollisional character of exchange narrowing. More discussion on the mechanism of the noncollisional (static) exchange narrowing in solutions can be found elsewhere [3.15].

The identical character of the dependences between ΔH and C for collisional and noncollisional mechanisms makes the measurement of spin exchange rate constants K_e from the exchange narrowing of ESR lines in concentrated solutions rather a complicated problem which in many cases cannot be solved.

Comparison of Spin Exchange Rate Constants Measured by Different Techniques

The comparison of the values of the spin exchange rate constants measured from the broadening, shift, and narrowing of ESR spectra is of importance not only for verifying the correctness of the measurement techniques, but

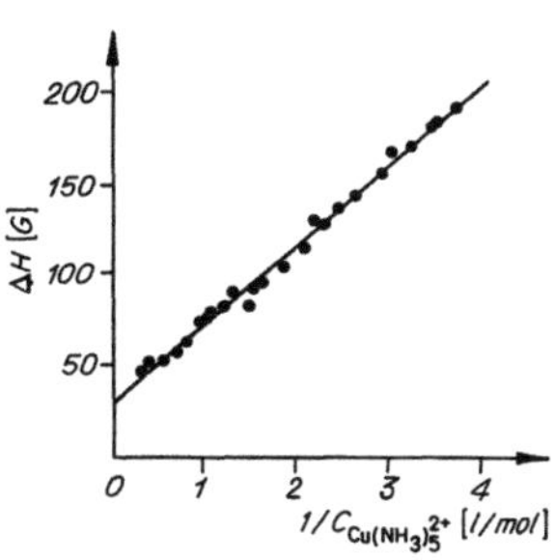

Fig.3.8. Concentration dependence of the line width for the copper-ammonia complex in the region of exchange narrowing [3.15]

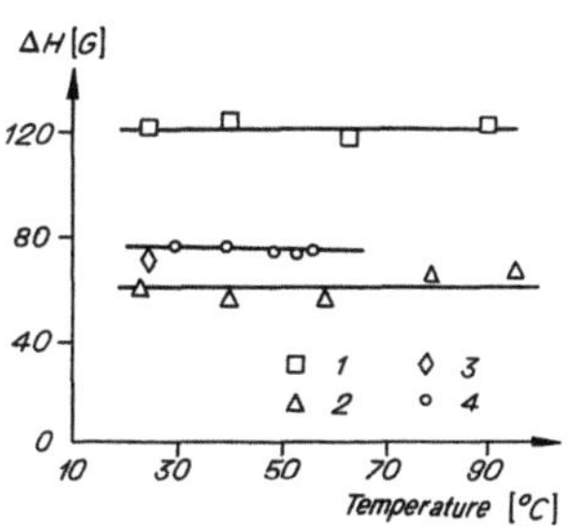

Fig.3.9. Influence of temperature and viscosity on the width of the exchange-narrowed line. Solution of $Cu(tmen)_2(H_2O)_2^{2+}$ in water (mol/l): (1) 0.8; (2) 1.8; solution of 2 mol/l of $Cu(en)_2(H_2O)_2^{2+}$ in water (3) and in a 1:4 mixture of glycerol and water (4)

also as a test of the validity of the theory (see Sect.2.4) describing the spectroscopic manifestations of spin exchange.

Unfortunately, because of numerous experimental difficulties (poor solubility or association of paramagnetic particles, influence of dipole-dipole interactions, change of solution viscosity with concentration, etc.) it was possible to measure the values of K_e simultaneously by several methods for only a few systems (Table 3.1).

Table 3.1. Comparison of K_e measured by different techniques

Paramagnetic particles	Solvent	$K_e \cdot 10^{-9}$ [l/mol·s]			References
		Broadening	Shift	Narrowing	
$(SO_3)_2NO^{2-}$	H_2O	1.72 [a]	1.42 [a]	-	[3.16]
$(CH_3)_3C\text{-}N(O\cdot)\text{-}C(CH_3)_3$	Dymethylformamide	3.7 ± 0.1	-	3.8 ± 0.3	[3.8]
$VOSO_4$	H_2O	5.5 ± 0.5	8 ± 4	-	[3.7]

[a] Rate constant K_e depends on concentration. The listed values of K_e are extrapolated to $C = 0$ and averaged over two series of measurements.

It is seen that within the limits of experimental errors the values of K_e determined by different procedures for one and the same system do coincide.

The systems listed in Table 3.1 can also be used to illustrate the difficulties arising in experiments on measuring K_e by various techniques. Thus, in the case of the anionradical $(SO_3)_2NO^{2-}$ the values of K_e could not be measured from an exchange narrowing of ESR lines seemingly due to the instability of this anionradical in concentrated solutions. In the case of the radical DTBN accurate measurements of exchange shifts are complicated by a comparatively large width of hyperfine lines which is due to unresolved additional hyperfine structure from protons. Finally, in the case of $VOSO_4$ the exchange narrowing occurs at such high concentrations that the static mechanism of exchange narrowing becomes operative.

3.1.2 Exchange Between Different Particles

In the case of exchange between different particles the rate constants are measured from an analysis of the ESR spectrum changes of one particle as the concentration of the other particle changes. The particle whose spectrum is analyzed is usually called the "indicator" while its partner is referred to as the "broadening agent".

For the spin exchange between different paramagnetic particles A and B the procedure for determining the rate constant is usually simpler than for exchange between identical particles. Particularly simple is the case when the spectrum of one of the partners – say, B – is not observed, e.g., due to a short relaxation time of the electron spin. In this case the spin exchange can only lead to the broadening of the line of particle A while practically no shift or narrowing is observed, as distinct from the case of exchange between identical particles. If the lines of the initial spectrum have a Lorentzian shape, their broadening is directly proportional to the concentration of particles B at any practically attainable C_B and is described by the formula

$$\Delta H = \Delta H_0 + (2/1.52 \cdot 10^7 g) K_e C_B \tag{3.30}$$

whence

$$K_e = 1.52 \cdot 10^7 (g/2)(\Delta H - \Delta H_0)/C_B \quad , \tag{3.31}$$

where g is the g value of particle A. Those formulas unlike the similar formula (3.21) for the exchange between identical particles contain no nuclear statistical factor since for the spin exchange between different particles it is always equal to unity.

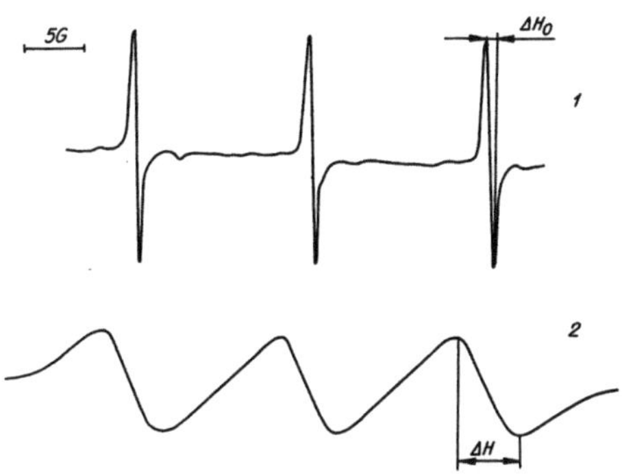

Fig.3.10. ESR spectra of TANONE radical in aqueous solution without additions (1) and with an addition of 0.17 mol/l of $CoClO_4$ (2) [3.17]

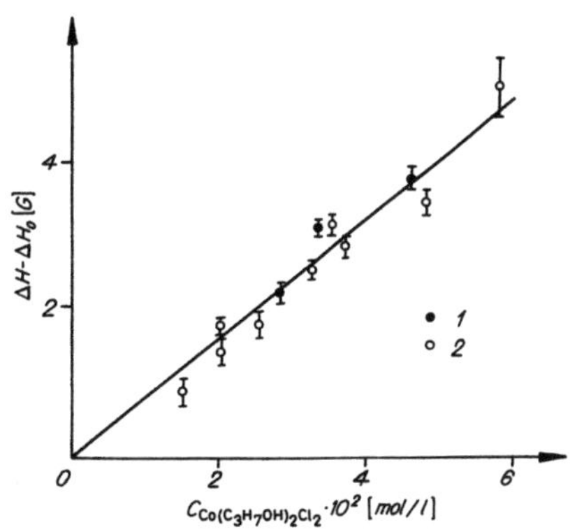

Fig.3.11. Broadening of the TANONE hfs component in the presence of $Co(C_3H_7OH)_2Cl_2$ [3.17]. (1) deoxygenated samples; (2) nondeoxygenated samples

As an example Fig.3.10 demonstrates the changes in the ESR spectrum of the free radical 2,2,6,6-tetramethyl-4-oxopiperidine-1-oxyl (TANONE) induced by the spin exchange with the aquacomplexes $Co(H_2O)_6^{2+}$. Figure 3.11 represents a broadening of the hyperfine component of the TANONE radical as a function of concentration of the complex $Co(C_3H_7OH)_2Cl_2$ in propanol for the range of concentrations in which the triplet hyperfine structure of the spectrum is well resolved. As is seen in accord with (3.30) the dependence of ΔH on C is linear within the range of cobalt concentrations $(1.5-5.75)\cdot 10^{-2}$ mol/l. From the data of Fig.3.11, one can calculate the rate constant $K_e = 1.0\cdot 10^9$ l/mol·s.

When the lines of both paramagnetic particles A and B are observed in the ESR spectrum the spin exchange, just as in the case of identical particles, may lead not only to a broadening but also to a shift of the lines and to a narrowing of the spectrum. However, the following peculiarity of the spin exchange between different particles should be mentioned. The total singlet line formed as a result of an exchange broadening and overlapping of the ESR hyperfine structure of particles A may, as the concentration of B particles increases, continue to broaden rather than to get narrower immediately after overlapping as in the case of exchange between identical particles until the spectra of A and B merge completely. Only after that does the narrowing of the total spectrum take place.

Until now the spin exchange rate constants have been determined experimentally for this kind of systems only from the broadening of hyperfine compo-

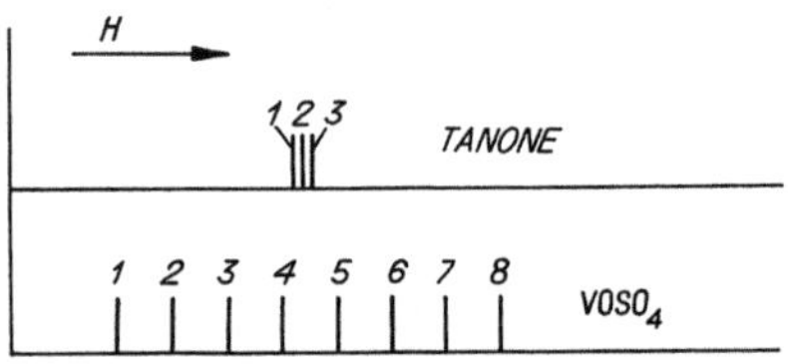

Fig.3.12. Schematic representation of the ESR spectrum for the TANONE + $VOSO_4$ system. The hyperfine structure in TANONE and $VOSO_4$-spectra originates from the interactions of the unpaired electrons with the nitrogen and vanadium nuclei, respectively

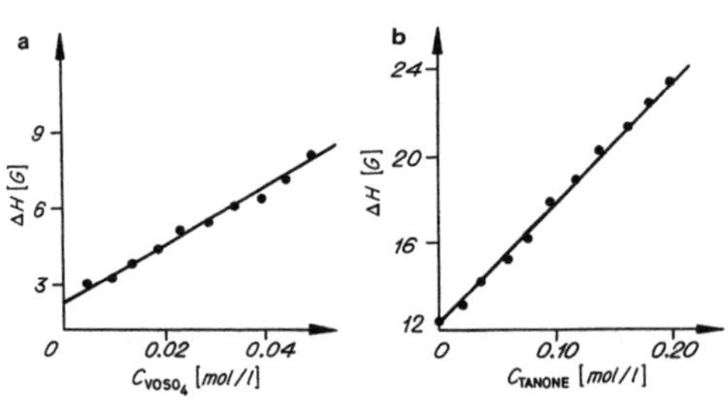

Fig.3.13a,b. Concentration broadening of component 2 of the ESR spectrum of TANONE in the presence of $VOSO_4$ (a) and of component 3 of the ESR spectrum of $VOSO_4$ in the presence of TANONE (b) [3.2]

nents (Figs.3.12,13). The procedure of finding K_e in this case does not differ from the one just described.

The spin exchange rate constants calculated from the data of Fig.3.13a,b are equal, respectively, to $(1.2 \pm 0.2) \cdot 10^9$ and $(0.8 \pm 0.2) \cdot 10^9$ l/mol·s, i.e., they practically coincide in agreement with the theory.

Just as in the case of exchange between identical particles, determination of K_e from exchange broadening becomes a more complicated problem if the lines of the initial spectrum in the absence of exchange have a nonzero initial width or a non-Lorentzian shape. It should be noted that in this situation the deviation from the Lorentzian shape may result not only from the presence of the unresolved hyperfine structure, as for exchange between identical particles, but also from the spin exchange between A particles or between B particles.

Let us consider the influence of the above factors in more detail. If the initial spectrum lines of A are Lorentzian, then a deviation from the linear dependence between their experimentally measured widths and the concentration of B takes place as soon as the total (initial + exchange) width of the lines of A comes up to 30-40% of the splitting between the lines. It should be noted that the spin exchange with the other (B) particles, unlike that between the same particles, in many cases only broadens the lines of A without changing their Lorentzian shape.

Such a situation arises, for example, when no ESR spectrum of B is observed or when it is sufficiently shifted from the studied spectral component of A. In this case the deviation of the concentration dependence of broadening from linearity results from the fact that, due to the overlapping of the adjacent lines of A, their experimentally measured width between the points of the maximum slope is different from the true one: $\Delta H_L = \Delta H_0 + \Delta H_e$. Thus, the task of measuring K_e is reduced merely to finding ΔH_L from the incompletely resolved spectrum of A particles. This can always be done by comparing the experimentally observed ESR spectra of A particles with the computer-simulated spectra consisting of a set of Lorentzian lines with the same ratio of the line intensities. The profiles of these theoretical spectra can be found in the Atlas of ESR spectra [3.4] or easily obtained by direct calculations. Let us consider as an example the procedure of determining K_e from the spectra of the bisdiphenylchromium cation (BDC) [3.17].

Curve 1 of Fig.3.14 represents the BDC ESR spectrum in a solution diluted to such an extent that the exchange between the BDC particles themselves no longer contributes to the line width, and curve 2 depicts the spectrum of the same solution after addition of $Co(ClO_4)_2$. The spectrum of BDC consists of 11

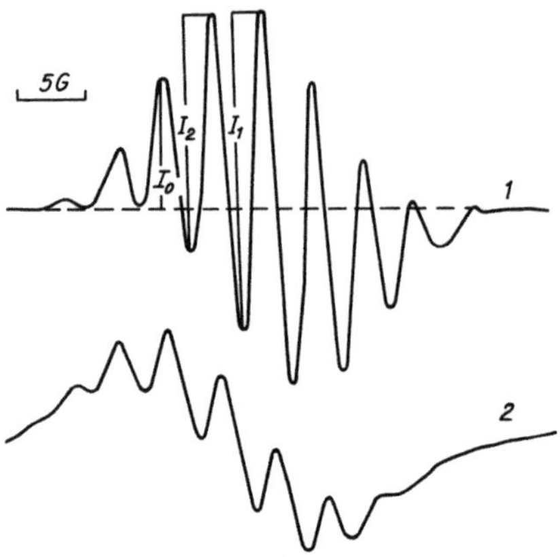

Fig.3.14. ESR spectra of the BDC cation without an addition (1) and with an addition of 0.15 mol/l of paramagnetic salt $Co(ClO_4)_2$ (2) [3.17]

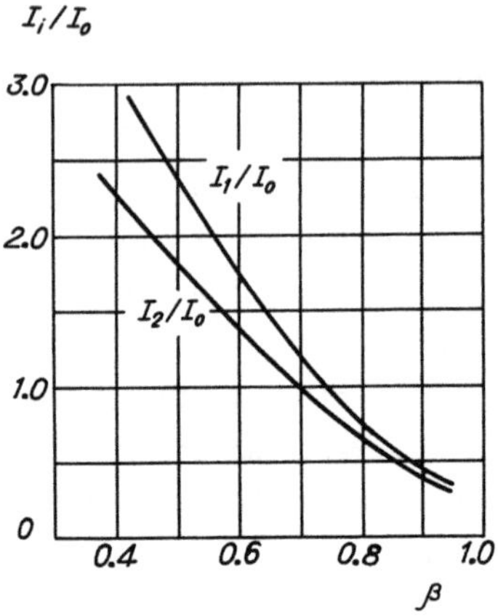

Fig.3.15. Calibration curve for determining the width of the hfs component of the spectrum with the ratio of intensities 1:10:45:120:210:252:... for the Lorentzian line shape [3.17]. The designations for I_1, I_0, I_2 are the same as in Figure 3.14

equidistant hyperfine components from 10 equivalent protons (the ratio of intensities is 1:10:45:120:210:252...; in Fig.3.14 the outer components cannot be seen because of their low intensity). In curve 2 of Fig.3.14 the lines are seen to be markedly broadened due to spin exchange with Co (II) complexes.

To determine this broadening, theoretically calculated spectra with the same ratio of integral intensities of the Lorentzian components were analyzed [3.17] to build calibration curves (Fig.3.15). With the help of these curves one can obtain a parameter $\beta = \Delta H_L/\Delta H_{hf}$ from the ratios of hyperfine component amplitudes which can be readily measured in experiment (see Fig.3.14, curve 1). The values of β can then be used to calculate the true line width of BDC $\Delta H_L = \beta \cdot \Delta H_{hf}$ (ΔH_{hf} for BDC equals 3.7 G) and the exchange broadening $\Delta H_e = \Delta H_L - (\Delta H_L)_0 = 2K_e C/g \cdot 1.52 \cdot 10^7$ for any concentrations of the broadening agent. An example of the dependence between the value of ΔH_L found with the help of this procedure and the concentration of the broadening agent is given in Fig.3.16. This dependence is indeed seen to be linear. The rate constant of the spin exchange between BDC and $CoPy_4Cl_2$ as calculated from the data of Fig.3.16 equals $2.6 \cdot 10^9$ l/mol·s.

If the lines of the initial spectrum of A do not overlap but their shape is non-Lorentzian due to the presence of an unresolved hyperfine structure, then the dependence of the experimentally observed width ΔH on the concentra-

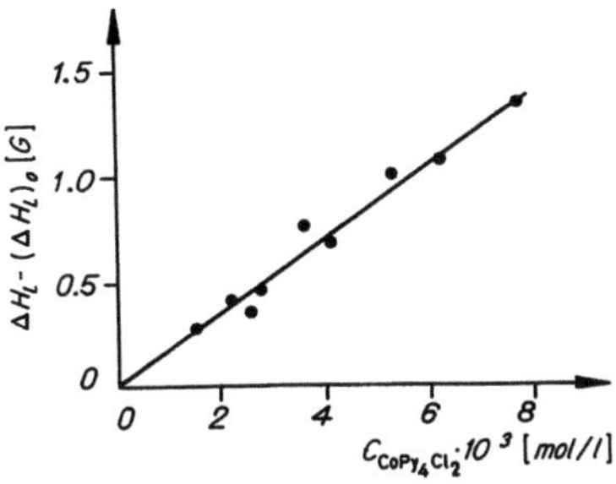

Fig.3.16. Broadening of the hfs components of the BDC cation ESR spectrum in a pyridine solution with an addition of $CoPy_4Cl_2$ [3.17]

tion of B at small exchange broadening ($\Delta H_e \lesssim \Delta H_0$) can also deviate from linearity just as in the case of exchange between identical particles.

In the particular, but practically important case of binomial distribution of the intensity of unresolved components their envelope is close to a line of Gaussian shape. The spin exchange with B particles induces the broadening of every unresolved component of the spectrum of A by the same value of $\Delta H_e = 2K_eC/g \cdot 1.52 \cdot 10^7$, their shape remaining Lorentzian. According to the results of numerical computations [3.6], the experimentally observed width ΔH between the points of the maximum slope for the spectrum as a whole is related to the width ΔH_G for the Gaussian profile approximating the unresolved hyperfine structure, and to the width ΔH_L for the individual Lorentzian components of the unresolved hyperfine structure as follows:

$$(\Delta H_G/\Delta H)^2 + \Delta H_L/\Delta H = 1 \quad . \qquad (3.32)$$

Hence the true width of Lorentzian components can be calculated from ΔH with the help of the formula

$$\Delta H_L = \Delta H[1 - (\Delta H_G/\Delta H)^2] \quad . \qquad (3.33)$$

It is seen from (3.32) or (3.33) that at $\Delta H \gg \Delta H_G$ the width $\Delta H \to \Delta H_L$. Therefore in practice, even in the presence of an unresolved hyperfine structure, the spin exchange rate constant can be determined directly from the slope of the concentration dependence of ΔH within its linear part just as was done in the case of spin exchange between TANONE and $Co(C_3H_7OH)_2Cl_2$ or exchange between identical particles. Note, however, that accurate measurement of K_e with the help of this procedure is possible only when the linear region of the curve is sufficiently long. Otherwise approximation of the curve by the linear dependence may result in too low a value of the experimentally measured rate constant since both in the region where the influence of unresolved hyperfine structures prevails (low C) and in the region where the influence of the overlap of resolved hyperfine components prevails (high C) the experimentally observed line width ΔH changes with the increase of concentration more slowly than does the Lorentzian width ΔH_L.

If the hyperfine lines of an experimantally observed ESR spectrum overlap and are neither Lorentzian nor Gaussian, then the reliable determination of K_e is only possible with the help of computers.

Consider as an example the measurement of the spin exchange rate constant from the broadening of the ESR lines of TANONE by $Cu(H_2O)_6^{2+}$ at high concentrations of TANONE. Under these conditions the exchange between the TANONE particles themselves distorts the shape of their ESR spectrum line [3.18].

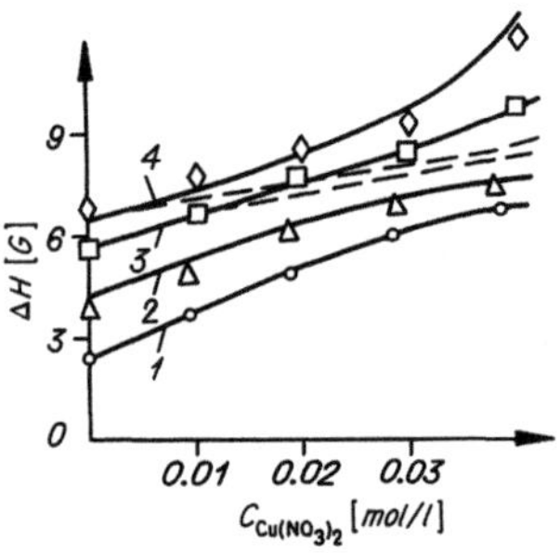

Fig.3.17. Dependence of the width of the central component of the TANONE ESR spectrum on the concentration of copper aquocomplex [3.18]. TANONE concentration (mol/l): (1) 0.01; (2) 0.03; (3) 0.05; (4) 0.06. (o,◇,□,△) - experimental data. (—) accurate computer calculations. (---) calculations using the Atlas [3.4]

Figure 3.17 represents the dependences of the width ΔH of the central hf component of TANONE on the concentration of $Cu(H_2O)_6^{2+}$ at various concentrations of TANONE. For $C_{TANONE} = 0.01$ mol/l the deviation of dependence from linearity at high $C_{Cu(NO_3)_2}$ is mainly due to the overlap of the adjacent hyperfine components of TANONE, while for $C_{TANONE} = 0.06$ mol/l it is mainly due to a non-Lorentzian shape of the TANONE spectral line. It is seen that for $C_{TANONE} = 0.06$ mol/l the experimental points do fit well the solid curve obtained by direct computations of the exchange-broadened line profiles and deviate from the dashed one obtained with the help of the Atlas [3.4] for three overlapping Lorentzian hyperfine components with the same integral intensity. The following table lists the rate constants of spin exchange between TANONE and $Cu(H_2O)_6^{2+}$ obtained from the initial linear parts of the $\Delta H = f(C)$ curves (see Fig.3.17) using (3.31). For comparison values obtained and with the help of computers, taking into account the corrections for the non-Lorentzian shape of the adjacent lines, are also listed.

C_{TANONE}, mol/l	0.01	0.03	0.05	0.06
$K_e \cdot 10^{-9}$, l/mol · s [using (3.31)]	2.09	1.65	1.46	1.41
$K_e \cdot 10^{-9}$, l/mol · s (using computer)	2.20	2.18	2.18	2.23

It is seen that the true values of K_e as expected are independent of C_{TANONE}. But the uncorrected values obtained with the help of the simple relation (3.31) depend on C_{TANONE}, and at large values of this parameter differ from the true ones by 30-40%.

In conclusion it should be noted that on the whole the procedure of calculating the spin exchange rate constants in the region of large broadenings for exchange between different particles is, as a rule, simpler than for exchange between identical ones. In practice this usually allows the values of K_e to be determined from a larger range of broadenings.

3.2 Other Methods of Measuring the Spin Exchange Rate Constants

The rate constants of spin exchange can be determined not only from ESR line broadening, shift, and narrowing but also from a number of other spectral properties such as saturation of ESR spectra in strong microwave fields, migration of spin excitation between various spectrum components in double resonance experiments, and decay of electron spin echo signals (see Sect.2.4). Because of their relative complexity these techniques have so far found much less application in studying spin exchange in solutions as compared to simpler methods described in the previous section. We shall, therefore, discuss them more briefly.

3.2.1 Continuous Wave Saturation Method

The saturation behavior of an individual ESR spectrum component depends on whether this component is a homo- or heterogeneously broadened line [3.19]. Heterogeneous broadening may be due to a number of cases, such as an unresolved hfs, to an overlap of lines of various paramagnetic species, etc. To establish the nature of line broadening is not a trivial problem. To choose between the two situations the analysis of the line shape can be of use. A homogeneously broadened line has a Lorentzian shape both in weak nonsaturating and in strong saturating microwave fields. A characteristic example of a Lorentzian line is the exchange-narrowed ESR spectrum [see (2.101)]. In contrast to that a heterogeneously broadened line is often well approximated by a Gaussian profile. An example is the envelope of the unresolved hfs from many magnetic nuclei. Such a situation is typical, e.g., of nitroxide radicals where the nitrogen components of the spectrum are heterogeneously broadened due to weak hf interaction with protons. For unambiguous judgment about the character of the line, however, the analysis of the line shape alone is not sufficient. Additional experiments are necessary (e.g., using the electron spin echo, see below [3.20]).

In the general case ESR spectra under the saturation conditions can be analyzed with the help of (2.70,100). For complicated spectra such an analysis is only feasible by computers. However in some cases of practical interest there are comparatively simple ways of analyzing a saturation effect. For example, there are convenient algorithms for analyzing the saturation of resolved spectra whose components are homogeneously broadened. Such a situation is rather typical of spin exchange experiments. Assume the spectrum of an isolated particle to consist of several components, each being heterogeneously broadened due to an unresolved hfs. Spin exchange induces spectral diffusion both between the components and inside each of them. Only at the lowest concentrations of paramagnetic particles when the spin exchange rate is also low enough will the spectrum components be saturated as heterogeneously broadened lines. As the spin exchange rate increases an exchange narrowing of the components to homogeneous lines will occur.

Consider now the saturation of a spectrum whose components are homogeneously broadened. As has been pointed out in Sect.2.4, the saturation effect yields additional information on spin exchange only in the region of the exchange broadening of spectrum components. In this situation the spectrum for any H_1 is given by (2.108). According to (2.108) each component of the spectrum is described by the Lorentzian function. In the case of saturation the lines get broader. The square of the line width between the points of the maximum slope for the kth component is equal to

$$\Delta H_k^2 = (4/3)[1/\gamma^2 T_2^2(k) + H_1^2 \cdot T_1(k)/T_2(k)] \quad , \tag{3.34}$$

where $T_2(k)$ and $T_1(k)$ are the time of transversal and the effective time of longitudinal relaxation of the kth component. In a typical experiment ΔH_k is usually measured at different amplitudes of the microwave field H_1 and then the dependence of ΔH_k^2 on H_1^2 is plotted. The intercept of the linear dependence thus obtained gives $4/3\gamma^2 T_2^2(k)$, and the slope gives $4T_1(k)/3T_2(k)$. From the two values one can find $T_1(k)$ for various components.

By measuring $T_1(k)$ at different concentrations of paramagnetic particles one can calculate the spin exchange rate constant with the help of (2.107). These measurements can also be used to verify the spin exchange theory which predicts [see (2.107)] that $T_1(k)$ must first decrease and then reach the plateau as the spin exchange rate increases. The asymptotic value of $T_1(k)$ must depend on the intensity of the component studied.

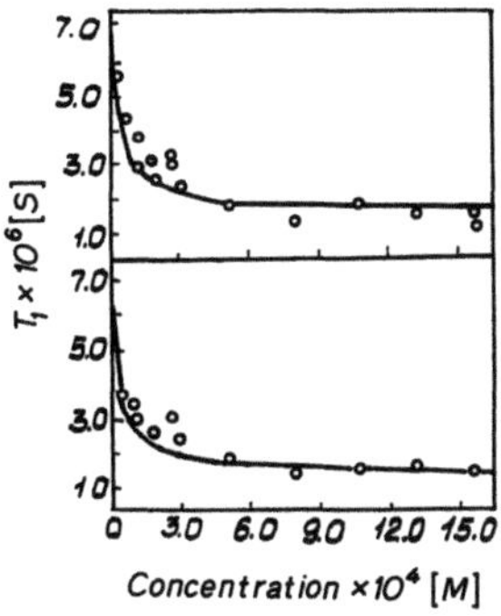

Fig.3.18. The concentration dependence of T_1 for $TCNE^-$ in DME at 15°C for two components of the spectrum. (o) represent the experimental data, (—) theoretical calculations [3.21]

Figure 3.18 presents theoretical calculations and experimental data obtained in the study of the ESR spectra saturation of $TCNE^-$ radicals [3.21]. The concentration dependence of $T_1(k)$ is seen to be in good agreement with the theory of spin exchange. According to (2.107)

$$[1 - T_1(k)/T_1]^{-1} = 1/[1 - \varphi(k)] + 1/[1 - \varphi(k)]K_eCT_1 \quad . \tag{3.35}$$

Thus in order to determine the spin exchange rate constant one should plot the dependence of $[1 - T_1(k)/T_1]^{-1}$ on $1/C$. The value of K_e can be determined from the slope of this plot if one knows the time T_1 of the spin-lattice relaxation of an isolated paramagnetic particle in the absence of exchange.

However the described procedure for determination of K_e from the analysis of the saturation effect in practice is often associated with a number of difficulties. A serious problem can be, e.g., the accurate measurement of the amplitude of the microwave field H_1 [3.21]. The value of the spin exchange rate constant is rather sensitive to the value of T_1, which must be measured in very diluted solutions, i.e., at $C \sim 10^{-5}$M [3.21] which is not so easy experimentally. Besides one has to measure small changes in the width of narrow lines with a small signal-to-noise ratio. Inaccurate knowledge of the quality of the spectrometer cavity in the presence of the sample and variations of the microwave field along the specimen lead to an additional 20-30% error in determining T_1 [3.21,22]. In principle, it is possible to measure T_1 directly from the kinetics of the spectrum recovery after a pulse saturation of the whole spectrum [3.23]. In this case in order to avoid the masking influence of spectral diffusion on the picture of spectrum recovery, all the spins must be saturated to the same degree which is difficult to do experimentally. Two additional important difficulties in determining the spin exchange rate constants from the saturation of ESR spectra should be mentioned. Firstly, spin exchange does not change much the effective time of spin-lattice relaxation [see (2.107) and Fig.3.18]. Thus the parameter which is measured in saturation experiments is often much less sensitive to the rate of spin exchange than, e.g., the ESR line width. Secondly, in the general case, the transfer of spin excitation between the hfi components may be induced not only by spin exchange, but by electron-nucleus transitions [3.21] and by dipole-dipole interaction between paramagnetic particles as well [3.24]. Only at relatively high concentrations of paramagnetic particles and high temperatures can one be sure that the decisive contribution to the process of spectral diffusion is made by spin exchange.

An interesting case is the spin exchange between free radicals and paramagnetic complexes with short paramagnetic relaxation times $T_{1,2} \sim 10^{-11}$ - 10^{-13}s [complexes of Co(II), Fe(II), Yb(III), etc.]. In relation to radicals these complexes act as components of the lattice. Thus the spin exchange between radicals and paramagnetic complexes makes the same contribution to the rates of transversal and longitudinal relaxation

$$\Delta(1/T_2) = \Delta(1/T_1) = K_{eA}C_B \quad , \tag{3.36}$$

where the spin exchange rate constant is given by (2.195,211). The contribution of spin exchange to the effect of saturation for this kind of solution — with two types of paramagnetic particles — was studied in [3.23,25].

3.2.2 Electron-Electron Double Resonance

The instruments and the different techniques of recording ELDOR spectra were developed by HYDE et al. [3.26,27]. The theoretical background of ELDOR application for spin exchange studies was briefly outlined in Sect.2.4. A detailed theory of ELDOR has been developed for radicals with one paramagnetic nucleus having the spin $I = 1/2$ and $I = 1$ and successfully applied for the first time to study spin exchange between nitroxide radicals [3.26]. A steady-state version of ELDOR was used for this purpose: one ESR spectrum component was saturated while for the other a decrease of the line intensity was observed. The ELDOR effect is characterized by a reduction parameter (2.109)

$$b = 1 - I/I_0 \quad ,$$

where I and I_0 are the intensities of the observed line with and without microwave pumping, respectively. Table 3.2 lists as an example the values of b for a 10^{-4}M solution of nitroxide radical in ethyl benzene [3.26].

With pumping one line the intensity of the other line may decrease via two quite different mechanisms: the transfer of excitation energy between the

Table 3.2. The values of factor b for nitroxide radical ($m=+1$ is the low-field, $m=0$ the central, and $m=-1$ the high-field component of the spectrum) [3.26]

Line of pumping	Observed line		
	+ 1	0	- 1
A) - 80°C			
+ 1	-	0.42	0.285
0	0.382	-	0.465
- 1	0.225	0.322	-
B) - 50°C			
+ 1	-	0.216	0.061
0	0.226	-	0.222
- 1	0.072	0.203	-
C) - 20°C			
+ 1	-	0.095	0.033
0	0.109	-	0.222
- 1	0.03	0.10	-
+ 10°C			
+ 1	-	0.030	0.019
0	0.034	-	0.031
- 1	0.017	0.024	-

pumped and the observed components and the direct saturation of the observed line on the wing of the pumped one.

The latter phenomenon complicates the study of the basic ELDOR effect associated with the transfer of excitation between the pumped and the observed components. One can, however, separate the two mechanisms if one studies the dependence of b on the amplitude of the pumping field. Direct pumping may decrease I down to zero ($b=1$) while the transfer of excitation decreases b only to a certain limit, $b<1$. Figure 3.19 [3.26] presents the dependence of b^{-1} on H_1^{-2} for a 10^{-4}M solution of nitroxide radical. It can be seen that as the power of pumping increases, b does reach a certain limiting value less than 1. This fact shows that in the present case the decrease in intensity of the line observed is due to the excitation transfer between the components. And it is only a further considerable increase in the pumping field amplitude that results in an additional decrease in the intensity of the observed line due to its direct saturation.

The ELDOR effect depends on the ratio of the rate of intercomponent transfer of excitation to that of spin-lattice relaxation of paramagnetic particles. For example, for radicals having one magnetic nucleus with spin 1/2 from (2.110) we have ($C_A=C/2$)

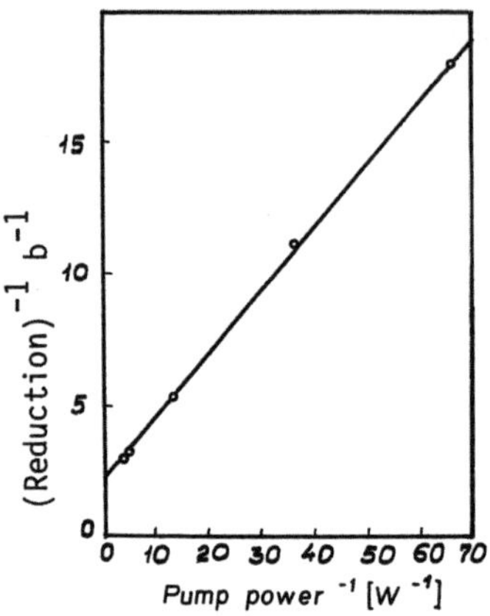

Fig.3.19. The dependence of b on the pumping power for a 10^{-4} M solution of nitroxide radical in ethylbenzene at - 80°C (the low-field spectrum component is observed while the central one is being pumped) [3.26]

$$b = K_eCT_1/(2 + K_eCT_1) \quad . \tag{3.37}$$

From an experimental value of b one can calculate K_eCT_1 and then K_e provided T_1 is known.

But there are a number of difficulties facing the explorer in practical implementation of this programm [3.26]. The theory of ELDOR leads to simple expressions of the type of (3.37) only in the case of complete saturation of the pumped line. Therefore, to find accurate values of the rates of relaxation processes one should measure the asymptotic value of b at large H_1 (see Fig.3.19). As the spin exchange rate grows the conditions for the saturation of the pumped line deteriorate. This may lower the accuracy of determining K_e. Besides, to find K_e from the ELDOR effect one should know T_1. In the above discussion of the steady-state saturation method it was emphasized that an accurate measurement of T_1 of isolated paramagnetic particles in solution is a difficult problem. And, finally, one should bear in mind that it is not the spin exchange alone that can produce the ELDOR effect. It has been shown, e.g., that at low temperatures and low concentrations of nitroxide radicals the main mechanism of spin excitation transfer between the spectrum components are electron-nucleus transitions induced by anisotropic hf interaction [3.26]. But at high temperatures and high concentrations of radicals the main mechanism producing the ELDOR effect is spin exchange.

3.2.3 Electron Spin Echo

To study spin exchange between radicals in solution use can also be made of the electron spin echo technique [3.20]. For this purpose the kinetics of decay of the primary and stimulated echo signals are studied. As seen from

(2.111,112) in the region of exchange broadening the spin exchange causes an exponential decay of echo signals. The dependence of decay rate on spin concentration and on temperature makes it possible to separate the contributions of spin exchange and of dipole-dipole interaction to paramagnetic relaxation.

The electron spin echo method is a useful complement to other methods of spin exchange study. For example, the analysis of exchange broadening, as well as of the saturation effects or ELDOR are rather time-consuming tasks for systems with unresolved or poorly resolved multicomponent ESR spectra. The electron spin echo method allows one to avoid the masking influence of the effects of heterogeneous broadening and to reveal the processes of irreversible longitudinal and transversal relaxation of spins. The application of this method in studying spin exchange is discussed in detail in [3.24]. So here we shall only note some of the difficulties associated with the use of this method. Modern electron spin echo spectrometers permit one to fully excite only those lines of the ESR spectra whose widths do not exceed 4-5 G. The broader spectra are only partially excited which seriously complicates the interpretation of spin echo data. Another problem concerns the resolution time of spectrometers. In modern instruments it amounts to tenths of a microsecond. Therefore in the region of exchange broadening the kinetics of echo signal decay can be really measured only for comparatively low spin exchange rates, when $K_eC < 10^7$ 1/s.

In conclusion it should be noted that though the methods described in this section are rather complicated they can sometimes provide us with information which is inaccessible via simpler techniques based on the analysis of the ESR line shapes. The values of spin exchange rates that can be measured by these more complicated methods are limited by the values of $1/T_1$ while for the methods based on the analysis of the line shape the limit is given by $1/T_2$. Since often $1/T_1 < 1/T_2$ the methods outlined in this section permit one in principle to measure smaller values of K_eC.

3.3 Separation of Exchange and Dipole-Dipole Contributions to Concentration Broadening of ESR Lines

As noted in the previous sections the most convenient and most frequently used method of spin exchange study in solutions is the measurement of the concentration broadening of ESR spectra. Such a broadening, however, may be induced not only by exchange but also by dipole-dipole interaction between paramagnetic particles. A correct measurement of spin exchange rate constants is only possible if the broadening due to dipole-dipole interactions is known exactly or is negligible compared with the broadening induced by exchange interactions.

The dipole-dipole contribution to the concentration broadening of ESR lines in solutions can be estimated from both direct experiments and theoretical calculations. In this section the results of such experiments and calculations are reported.

Theoretical Estimation of Dipole Broadening

The contribution of dipole-dipole interaction to the concentration broadening of lines depends on the solution viscosity since this interaction is averaged by diffusion of spins. The characteristic time of translational diffusion is the lifetime of two partners in the cage [3.28]

$$\tau_D = R_{AB}^2/D_{AB} \quad . \tag{3.38}$$

In order to estimate the order of magnitude of τ_D we assume the radii of the particles to be equal to $R_A = R_B = R_{AB}/2$, and express the diffusion coefficient through the viscosity using the Stokes equation. We obtain

$$\tau_D = 3\pi R_{AB}^3 \eta/2kT \quad . \tag{3.39}$$

For example, at $R_{AB} = 6\,\text{Å}$, $T = 25\,^\circ\text{C}$ (3.39) gives $\tau_D = 2.5\cdot 10^{-8}\eta$. Hence, e.g., for water solutions at room temperature ($\eta \approx 0.01$), we have $\tau_D = 2\cdot 10^{-10}$ s.

From the standpoint of the efficiency of averaging the dipole-dipole interaction the solution is viscous if

$$\tau_D^2 < \omega_d^2 > > 1 \quad , \tag{3.40}$$

where $< \omega_d^2 >$ is the second moment of dipole interaction (see [3.28])

$$< \omega_d^2 > = 5.1\gamma^4\hbar^2 S(S + 1)CR_{AB}^{-3} \quad , \tag{3.41}$$

where C is the concentration of paramagnetic particles. Substituting $\gamma = 1.76 \cdot 10^7$ rad/s·G and $R_{AB} = 6\,\text{Å}$ we obtain

$$< \omega_d^2 > = 10^{18} S(S + 1)C\ [\text{rad/s}]^2 \quad , \tag{3.42}$$

where C is measured in mol/l. As a result we arrive at the following condition for solutions to be viscous

$$C\eta^2 S(S + 1) \gtrsim 10^{-3} \quad , \tag{3.43}$$

where C is measured in mol/l and η in poises. For example, a water solution of radicals is to be considered as viscous from the point of view of averaging the dipole interaction at $C \gtrsim 10M$.

In viscous solutions the dipole broadening reaches its maximum possible value and becomes

$$\Delta H = (16\pi^2/27) < |\mu_z| > C \quad , \tag{3.44}$$

where $<|\mu_z|> = g\beta <|m|>$ is the average absolute value of the longitudinal magnetic moment component of the broadening agent, $<|m|> = (1+2S)/4$ for half-integral values of S and $<|m|> = S(S+1)/(1+2S)$ for integral values of S. Hence we obtain the maximum possible constant of dipole broadening

$$K_d(\max) = (8\pi^2/9\sqrt{3})\gamma < |\mu_z| > \quad . \tag{3.45}$$

For pure spin magnetism

$$K_d(\max) = 9.8 \cdot 10^8 < |m| > \quad [1/\text{mol} \cdot \text{s}] \quad . \tag{3.46}$$

For example, for free radicals

$$K_d(\max) = 4.9 \cdot 10^8 \quad [1/\text{mol} \cdot \text{s}] \quad . \tag{3.47}$$

The maximum constant of dipole broadening increases with increase in the spin of the broadening agent. When passing from the spin $S = 1/2$ to $S = 5/2$ $K_d(\max)$ increases threefold.

Even in viscous solutions the actual dipole broadening may be less than the maximum value (3.44) as a result of averaging the dipole-dipole interaction by spin flip-flops induced by spin-lattice relaxation, spin diffusion, etc. Such an averaging of dipole interactions in solids is discussed in detail elsewhere [3.24]. The dipole profile of the line in this situation is no longer Lorentzian; in the center the line becomes narrower while its wings fall off more slowly than for the Lorentzian shape. The theory of dipole-dipole broadening of the ESR lines in nonviscous solutions when $\tau_d^2 \cdot <\omega_d^2> < 1$ was developed by a number of authors [3.28-31]. In this case the Brownian motion of spins effectively averages the dipole-dipole interaction. The magnitude of the dipole broadening is determined by the spectral densities, $J(\omega)$, of the correlation function of the dipole-dipole interaction of two spins. Under the conditions of ESR experiments $J(\omega_0) < J(0)$ (the resonance frequency $\omega_0 \approx 6 \cdot 10^{10}$ rad/s) since at $\eta \gtrsim 0.1$ cP $\omega_0\tau_D > 1$. The dipole broadening of the kth component of hfs in solutions of identical paramagnetic particles is characterized by the constant

$$K_d^{(k)} = (1/6)\gamma^4\hbar^2 S(S+1)[5/4 + \varphi(k)]J^{(0)}(0) \quad , \tag{3.48}$$

where $\varphi(k)$ is the ratio of the intensity of the kth hfs component of the spectrum to the total intensity of all components. If the broadening agent is a particle of another kind (B) then

$$K_d = (5/24)\gamma^2\gamma_B^2\hbar^2 S_B(S_B+1)J^{(0)}(0) \quad . \tag{3.49}$$

If the hfs components of the spectrum are heterogeneously broadened then the dipole broadening of all components of the spectrum must be calculated by using (3.49) instead of (3.48) since in this case for any spin practically all the other spins are nonresonant.

The most accurate calculations of the spectral densities $J(\omega)$ are made in [3.30,31]. It is shown that

$$J^{(0)}(0) = (128\pi/135R_{AB} \cdot D_{AB})(1 + 3 < \lambda^2 > /8R_{AB}^2) \quad , \tag{3.50}$$

where $<\lambda^2>$ is the mean-root-square displacement of particles in an elementary step of diffusion. In the extreme case of continuous diffusion

$$J^{(0)}(0) = (128\pi/135)\tau_D/R_{AB}^3 \quad . \tag{3.51}$$

Using Stokes law and expressing the diffusion coefficient through viscosity from (3.48,49,51) we obtain

$$K_d^{(k)} = (32\pi^2/135)\gamma^4\hbar^2 S(S+1)[5/4 + \varphi(k)]\eta/kT \quad , \tag{3.52}$$

$$K_d = (8\pi^2/27)\gamma^2\gamma_B^2\hbar^2 S_B(S_B+1)\eta/kT \quad . \tag{3.53}$$

The numerical calculations of $J(\omega)$ show [3.30] that for paramagnetic ions with opposite charges equal to unity $J(0)$ increases in a water solution due to Coulomb attraction about twofold. In the case of repulsion for like-charged partners it is decreased about twofold as compared to (3.44). The calculations have been made within the framework of the Debye-Hückel theory for the temperature of 25°C.

Let us estimate K_d using (3.53). Considering the magnetism of the particles to be of purely spin nature ($g = 2.0023$) from (3.53) we obtain at $T = 25°C$

$$K_d = 4.5 \cdot 10^9 S_B(S_B+1)\eta \quad [1/\text{mol} \cdot \text{s}] \quad . \tag{3.54}$$

Hence, e.g., for water solutions, we obtain the estimate

$$K_d = 4.0 \cdot 10^7 S_B(S_B+1) \quad [1/\text{mol} \cdot \text{s}] \quad . \tag{3.55}$$

The complexes of such paramagnetic ions as Co^{2+}, Fe^{2+}, and most lanthanide ions have such short times of paramagnetic relaxation ($T_{1,2} < 10^{-11}$ s) that $T_{1,2} < \tau_D$ and $\omega_0 \cdot T_{1,2} < 1$. If such complexes are used as broadening agents, then the main mechanism of averaging the dipole-dipole interaction is the paramagnetic relaxation of spins rather than the translational diffusion, and the constant of dipole broadening becomes equal to

$$K_d = (5/6)\gamma^2\gamma_B^2\hbar^2 S_B(S_B + 1)J^{(0)}(0)$$

$$= (16\pi/9)\gamma^2\gamma_B^2\hbar^2 S_B(S_B + 1)T_{1B}R_{AB}^{-3} \quad . \tag{3.56}$$

For example, for $R_{AB} = 6\text{Å}$ and $T_{1B} = 10^{-12}$ s we have

$$K_d = 1.6 \cdot 10^6 S_B(S_B + 1) \quad [1/\text{mol} \cdot \text{s}] \quad . \tag{3.57}$$

The spectral densities $J(\omega)$ for nonviscous solutions at arbitrary paramagnetic relaxation times of the broadening agents have been calculated in [3.30,31].

Unfortunately, at present there is no theory of dipole broadening for the intermediate values of viscosity that are often of practical interest. One can reliably estimate only the upper extreme value of dipole broadening in viscous and nonviscous solutions. In the latter a significant averaging of the dipole interactions occurs. Comparison of (3.46) with (3.55) and (3.57) shows that in typical experimental conditions the dipole broadening may be 1 or 2 orders of magnitude smaller than its upper extreme value.

Experimental Estimation of Dipole-Dipole Contribution at Low Viscosity

The validity of estimates by means of (3.53) can be checked experimentally. Indeed, the value of K_d must be maximal for systems in which the broadening agents are the high-spin Mn(II) complexes with $S = 5/2$, the maximum value of S for metals with the unfilled d shell. The broadening constant $K_b = K_d + (5/6) K_e = 3.3 \cdot 10^8$ 1/mol·s have been found for the interaction of $Mn(H_2O)_6^{2+}$ complexes in water solution [3.32]. The factor 5/6 in front of K_e is the statistical factor $1 - \varphi$ [see (3.60)] for particles with the nuclear spin $I = 5/2$. Since the concentration broadening of the $Mn(H_2O)_6^{2+}$ line increases rather than decreases with the increase in viscosity, the main contribution to K_b is assumed to be made by K_d. In any case $K_d \leq K_b = 3.3 \cdot 10^8$ 1/mol·s. The value obtained agrees with the above estimate according to (3.53): $K_d = 3.9 \cdot 10^8$ 1/mol·s for $S = 5/2$ at $\eta = 1$ cP. The value of K_d is much less than that of K_d(max) in the case of a solid state [$1.5 \cdot 10^9$ 1/mol·s for $S = 5/2$ according to (3.45)]. Hence even for $S = 5/2$, and the more so for smaller S, the dipole-dipole interaction is effectively averaged by thermal motion. In these cases the value of K_d must be proportional to $S(S+1)$ where S is the spin of the broadening agent [see (3.53)]. Thus, for systems in which the broadening agent has $S = 3/2$ and $S = 1/2$ and size close to that of $Mn(H_2O)_6^{2+}$ we obtain, e.g., $K_d(S = 3/2) = (3/7) \cdot K_d(S = 5/2) = 1.4 \cdot 10^8$ l/mol·s and $K_d(S = 1/2) = (3/35) K_d(S = 5/2) = 2.8 \cdot 10^7$ l/mol·s [cf with the estimation $K_d = 3.4 \cdot 10^7$ l/mol·s according to (3.53) at $\eta = 1$ cP].

The value of K_d for systems with $S = 1/2$ may also be estimated from the data [3.8] on the broadening of the singlet exchange-narrowed line of the ESR spectrum of DTBN in concentrated solutions (see Fig.3.7) in dimethyl formamide ($\eta = 0.8$ cP). As has been already pointed out, this broadening is most likely due to the dipole-dipole interaction. From the slope of the dependence of ΔH on C_{DTBN} at $C_{DTBN} > 1$ mol/l one can calculate $K_d \approx 1.4 \cdot 10^7$ l/mol·s [cf with $K_d = 2.7 \cdot 10^7$ l/mol·s at $\eta = 0.8$ cP according to the estimate by means of (3.53)]. In this case the experimental value of K_d is about twice as small as that calculated theoretically. It is possibly due to an additional averaging of dipole-dipole interaction by spin exchange in concentrated solutions. The influence of this factor is not taken into account by (3.53). In the other cases the experimental estimates of K_d, as shown above, are in better agreement with the theoretical ones, but always K_d (theor.) > K_d (exper.). This seems to be also due to additional sources of averaging of the dipole-dipole interaction, such as spin exchange and spin-lattice relaxation. Consequently, the values of K_d obtained with the help of (3.53) should be regarded as the upper limit of the contribution of dipole-dipole interactions to the broadening constant at low viscosity.

Thus, on the basis of the above data one can state that in nonviscous ($\eta \lesssim 1$ cP) liquids the constants of broadening may a priori be regarded as practically coinciding with K_e (exchange between different particles) or with $(1 - \varphi) K_e$ (exchange between identical particles) if they exceed 10^8 l/mol·s for the broadening agents with $S = 1/2$, $3 \cdot 10^8$ l/mol·s with $S = 3/2$ and 10^9 l/mol·s with $S = 5/2$. To determine K_e at lower values of K_b it is necessary to estimate the contribution of K_d to K_b using the formula of the previous subsection or the experimental techniques given below.

For broadening agents with short relaxation times ($T_1 < \tau_c$) the dipole-dipole contribution to K_b becomes especially small. For such agents even smaller values of K_e can be measured reliably from concentration broadening. For example, for Co(II) complexes with $T_1 \sim 10^{-12}$ s and $S_B = 3/2$ the dipole-dipole broadening constant K_d according to (3.57) equals $6 \cdot 10^6$ l/mol·s. Hence for these complexes the values of $K_b \gtrsim 2\cdot 10^7$ l/mol·s can be already regarded as coinciding with K_e.

Separation of the Cases of Exchange and Dipole-Dipole Broadening According to Viscosity and Temperature Dependences

According to the theory presented in the beginning of this section the dipole-dipole contribution ΔH_d to the ESR line width in a liquid must grow with the growth of the factor η/T (η is liquid viscosity, T, temperature) which is a quantitative measure of the influence of medium on the rate of diffusion motion (see Sect.4.1).

The dependence of the spin exchange contribution ΔH_e to the line width on η and T in the general case is complicated, but for strong collisions this contribution must be proportional to T/η. Thus, in the case of strong collisions the dependences of dipole-dipole and exchange contributions to the broadening of ESR lines on the parameter η/T must follow opposite trends which allows one to distinguish between them.

Consider as an example the data [3.33,34] on the influence of viscosity and temperature on the broadening of the ESR lines of free nitroxide radicals, acting as indicators, when various paramagnetic complexes and radicals are added to the solutions. The solvents used were water-glycerol mixtures whose factor η/T was varied in some cases by changing the composition of the mixture at a constant temperature of 25°C, in others, by changing the temperature with the mixture composition unchanged. The solution viscosity varied from 1 to 400 cP. To estimate the possible concentration broadenings due to factors other than the paramagnetism of the added particle, control experiments were carried out with the diamagnetic salt $Cd(NO_3)_2$. The complexes and radicals used as broadening agents, the ranges of their concentration, as well as the indicators used and their concentrations are listed in Table 3.3.

The ESR line width ΔH of the indicator for all the systems studied was found to increase as a linear function of concentration of the broadening paramagnetic agent: the slope of the straight line $\Delta H = f(C)$ is related to the broadening constant K_b as [cf with (3.30)]

$$\Delta H = \Delta H_0 + 2\ K_b C/1.52 \cdot 10^7 g \quad , \tag{3.58}$$

where ΔH_0 is a constant concentration-independent contribution to the line width (e.g., due to spin-lattice relaxation).

The general character of the dependence K_b on η/T is the same for all the systems studied. As an example Fig.3.20 represents the dependences of K_b =

Table 3.3. ESR line broadening constants of indicators in water (K_b) and in water-glycerol mixture [K_d(max)] [3.33]

Indicator	Concentration of indicator [mol/l]	Broadening agent[a]	Range of concentrations [mol/l]	T_1[s] (aqueous solution)	$K_b \cdot 10^{-8}$ [l/mol · s] (η = 1 cP)	$K_d(max) \cdot 10^{-8}$ [l/mol · s]
R_2[b]	-	R_2	10^{-3} - 10^{-1}	$1 \cdot 10^{-6}$ [d]	13.3 ± 0.7	3.4 ± 0.7
R_1[c]	$5 \cdot 10^{-3}$	$Cu(en)_2(H_2O)_2^{2+}$	0 - $5 \cdot 10^{-2}$	$2.7 \cdot 10^{-9}$	25.9 ± 0.7	3.0 ± 0.7
R_1	$1 \cdot 10^{-2}$	$Mn(H_2O)_6^{2+}$	0 - $3 \cdot 10^{-2}$	$2 \cdot 10^{-9}$	20.5 ± 0.7	6.5 ± 1.5
R_1	$5 \cdot 10^{-3}$	$Cr(H_2O)_6^{3+}$	0 - $5 \cdot 10^{-2}$	$1 \cdot 10^{-9}$	12.3 ± 0.7	< 0.7
R_1	$5 \cdot 10^{-3}$	$Ni(en)_2(H_2O)_2^{2+}$	0 - $5 \cdot 10^{-2}$	$4.3 \cdot 10^{-12}$ [e]	21.0 ± 0.7	< 0.7
R_1	$5 \cdot 10^{-3}$	$Co(H_2O)_6^{2+}$	0 - $5 \cdot 10^{-2}$	$7 \cdot 10^{-13}$	8.0 ± 0.7	0.30 ± 0.05
R_1	$5 \cdot 10^{-3}$	$Cd(H_2O)_6^{2+}$	-	-	-	0.14 ± 0.09 [f]

a Nitrate salts were used in all cases.
b R_2 = 2,2,6,6-tetramethyl-4-oxypiperidine-1-oxyl (TANANE)

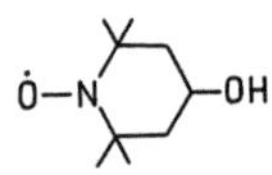

c R_1 = 2,2,6,6-tetramethyl-4-oxopiperidine-1-oxyl (TANONE)

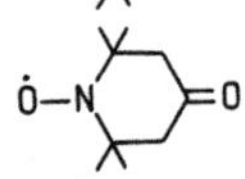

d Estimated from the line width.
e For aquocomplex.
f Measurements of K_d(max) were made in glycerol with 1-2 M Cd^{2+} concentration.

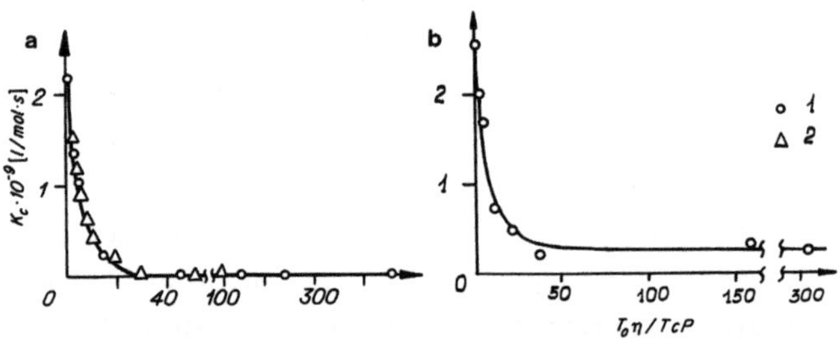

Fig.3.20. Dependence of the broadening constant for the systems R_1 + complex on the parameter $T_0\eta/T$ [3.33,34]. (a) The complex $Ni(en)_2(H_2O)_2^{2+}$; (b) the complex $Cu(en)_2(H_2O)_2^{2+}$ [(1) at a constant temperature of 25°C; (2) at a constant composition of the mixture]

$f(T_0\eta/T)$ (T_0 = 298 K) for the systems with the complexes $Ni(en)_2(H_2O)_2^{2+}$ and $Cu(en)_2(H_2O)_2^{2+}$ as the broadening agents.

In the case of $Ni(en)_2(H_2O)_2^{2+}$ the viscosity of solutions was varied in two ways: by changing the concentration of glycerol at constant temperature and by changing the temperature with the compositon of the water-glycerol mixture unchanged. As seen from Fig.3.20a, all the experimental points fit one curve. This seems to indicate that the changes in the solvation of reagents as the mixture composition changes and the possible microheterogeneity of the water-glycerol mixture do not essentially affect the line broadening. The curve of $K_b = f(T_0\eta/T)$ in the studied range of η/T variation is sufficiently well described by a hyperbolic dependence. Such a behavior of the curve corresponds to the change in the number of binary collisions in solutions and hence to the broadening caused by spin exchange in the situation of strong collisions.

In the experiments with $Cu(en)_2(H_2O)_2^{2+}$ the viscosity was varied by changing the concentration of glycerol at the constant temperature of 25°C. As seen from Fig.3.20b, at the viscosity of 1 to 50 cP the value of K_b also decreases with increasing η. As the viscosity grows further the broadening constant remains unchanged within the accuracy of the measurements. Such a behavior of the curve of $K_b = f(T_0\eta/T)$ can be accounted for if the observed ESR line broadening of the indicator is assumed to be induced by both exchange and dipole-dipole interactions with the broadening agent. The constant of broadening also seems to consist of two parts

$$K_b = K_e + K_d \quad ^2 \quad . \tag{3.59}$$

As the solution viscosity increases, $K_e \to 0$ and K_d increases and tends to the value $K_d(\max)$ corresponding to a solid. As is seen from a simple estimation with the relation $K_e \lesssim K_b(1\ \text{cP})/\eta$, where $K_b(1\ \text{cP})$ is the broadening constant observed experimentally at $\eta = 1$ cP, the contribution of exchange interactions to the constant of broadening at $\eta \approx 50$ cP must indeed be negligible. Thus, at high viscosities $K_b = K_d(\max)$, i.e., observed broadening is determined solely by dipole-dipole interaction. As is seen from Table 3.3, for a given indicator this quantity depends on the nature of the broadening agent.

It is evident that $K_d(\max)$ represents the upper limit for the values of K_d in a nonviscous liquid. As is seen from Table 3.3 for all the systems studied the constants of broadening at $\eta = 1$ cP are much larger than the values of $K_d(\max)$. Hence it can be concluded that at low viscosities ($\eta \lesssim 1$ cP) the broadening of ESR lines is practically totally determined by exchange interactions.

---

2 This relation holds only for the interaction of different particles. For that of identical ones it is to be replaced by

$$K_b = (1 - \varphi)K_e + K_d \quad , \tag{3.60}$$

where φ is the statistical weight of a line whose broadening is being measured, see (3.22).

From the data discussed in this section it follows that in all the examples considered in Sect.3.1 it is the exchange, rather than dipole-dipole, interaction that makes the main contribution to the broadening of ESR lines.

3.4 Separation of the Cases of Weak and Strong Exchange

To obtain information of interest for chemistry, biology, or molecular physics from spin exchange rate constants one should first answer the question: which kind of exchange – weak or strong – does the rate constant correspond to? Often this question cannot be answered a priori, on the basis of the data on the electron structure and diffusion mobility of paramagnetic particles. Thus the experimental ways [3.17] of choosing between the two situations must be considered.

As has been said above (see Sect.2.5) the rate constant in the region of strong exchange is given by the formula

$$K_e = fp_{max} \cdot K_D \quad , \tag{3.61}$$

where K_D is the rate constant of diffusion binary collisions, f the steric factor, and p_{max} the maximum value of spin exchange efficiency. In the case of weak exchange there appears an additional dependence of the spin exchange rate constant on the exchange interaction energy J, the collision time τ_c, and the relaxation time T_1 of the electron spin (if for the broadening agent $T_1 < \tau_c$). In the extreme case of the weak exchange between a radical ($S = 1/2$) and a paramagnetic particle with an arbitrary spin S, in accordance with the results of Sect.2.5, we have

$$K_e = fpK_D \cong (2/3)S(S+1)J^2\tau_c^2 fK_D \tag{3.62}$$
$$\text{if } \tau_c < T_1 \quad ,$$

$$K_e \cong (2/3)S(S+1)J^2T_1\tau_c fK_D \tag{3.63}$$
$$\text{if } \tau_c > T_1 \quad .$$

Since in these formulas $p < p_{max}$, it is the small value of the spin exchange rate constant that may serve as a certain argument in favor of weak exchange. It should be borne in mind, however, that constant K_e may also be small in the case of strong exchange if the latter is hindered sterically and due to this the factor f in (3.61) is small. Therefore the experimental methods of separating the strong and weak exchange are usually based not on the absolute values of rate constants but on their different dependences on the solvent viscosity η, the temperature T, the electron spin relaxation time T_1 of the broadening agent, and the value of the exchange integral J which is determined by the spin densities ρ on the partners' atoms accessible in collisions.

The criterion based on the dependence of K_e on T_1 and ρ is especially descriptive since such a dependence must be observed only in the case of weak exchange. This criterion can be conveniently applied in comparing various compounds of the same type since in this case it is a simpler matter to have the change of K_e be mainly determined by that of the chosen parameter and depend comparatively little on other factors. Thus in studying the spin exchange of free radicals with paramagnetic metal complexes it is easy to vary T_1 by many orders of magnitude replacing one metal atom by another in a series of similar complexes. This series can be so selected that the steric factor is expected to remain practically unchanged. Under these conditions, as is seen from (3.63), a decrease in K_e as T_1 decreases testifies in favor of weak exchange. Note that in such experiments the value of spin density on the ligands may change simultaneously because it can depend on the nature of the central atom as well. This circumstance, though it may complicate the dependence of K_e on T_1, is not in most cases an obstacle to the application of the criterion in question.

In the system "radical-paramagnetic complex" it is also rather simple to vary in a wide range (by several orders of magnitude) the spin density ρ on the external atoms of the complex by changing the ligands. The considerable change of K_e under these conditions may also testify to weak exchange though not always. It should be kept in mind that an increase in the size of the ligands may in fact result in a drop of K_e not only because of the decrease in spin density but also because of the steric screening of the complex's internal atoms whose spin density is high. In the latter case the spin exchange may remain strong. Thus, a decrease in K_e after bulky ligands are introduced cannot be an unambiguous argument for the exchange to become weak.

The criterion that is currently the most widespread due to its simplicity and reliability is one based on measuring K_e at different viscosities. Viscosity may be varied in different ways: by changing the temperature, solvent composition, or external pressure. The parameters in (3.61-63) most sensitive to changes in viscosity are K_D and τ_c: $K_D \sim \eta^{-1}$, $\tau_c \sim \eta$ (for more details see Sects.4.1,2). Taking into account these relationships and assuming the parameters J and f to be independent of viscosity and temperature we have for the strong exchange

$$K_e \sim T/\eta \quad . \tag{3.64}$$

For the weak exchange, depending on the relation between τ_c and T_1, we have

$$K_e \sim \eta/T \quad , (\tau_c < T_1) \quad , \tag{3.65}$$

or

$$K_e \sim T_1 \quad , (\tau_c > T_{1,2}) \quad . \tag{3.66}$$

As follows from comparing (3.64) with (3.65), for particles with long relaxation times the dependences of spin exchange rate constants on parameter T/η in the two situations are opposite. For particles with short relaxation times (3.64,66) also predict different dependences on T/η (Fig.3.21). Note that for the two situations $\tau_c < T_1$ and $\tau_c > T_1$ the character of the curves is the same in the region of the strong exchange but it is essentially different for the weak exchange. In the region of weak exchange K_e drops with the rise of T/η if $\tau_c < T_1$, and remains practically unchanged if $\tau_c > T_1$. Note that it is the dependence of the latter type, i.e., the invariability of the rate constant in the kinetic region of the reaction, that is characteristic of chemical reactions, while a decrease in the constant as T/η increases is peculiar to spin exchange.

If the viscosity is varied by changing the temperature, an alternative way of analyzing the experimental data is to plot the Arrhenius temperature depen-

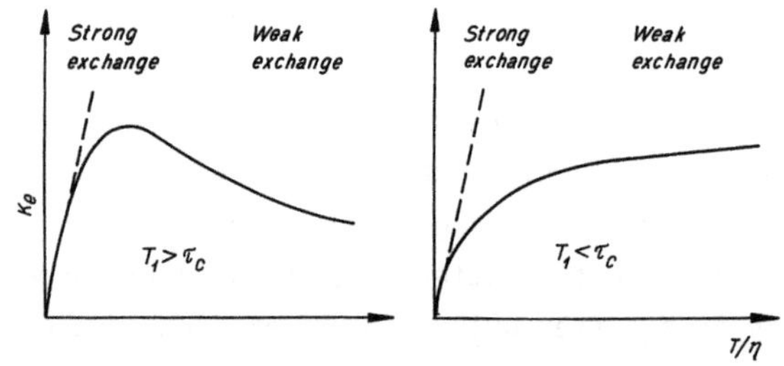

Fig.3.21. General view of the dependence of the spin exchange rate constant on T/η for long and short relaxation times of the partner. T_1 is assumed to be independent of T/η

dence of the exchange rate constant

$$K_e = K_{e0} \exp(-E_e/kT) \quad .$$

Extrapolating the temperature dependence of η also by the Arrhenius law

$$\eta = \eta_0 \exp(E_\eta/kT) \qquad (3.67)$$

and assuming in agreement with the experimental data T_1 to be relatively weakly dependent on temperature, we obtain the following expression for the activation energy of K_e:

strong exchange

$$E_e = E_\eta + RT \qquad (3.68)$$

weak exchange ($T_1 > \tau_c$)

$$E_e = - E_\eta - RT \qquad (3.69)$$

weak exchange ($T_1 < \tau_c$)

$$E^e = - RT^2(d/dT)\ln T_1 \approx 0 \quad . \qquad (3.70)$$

The possible types of the temperature dependences of K_e in the Arrhenius coordinates are shown in Fig.3.22. As is seen, the transition from strong to weak exchange may be accompanied by the change of the sign of the activation energy. If, as is often the case, T_1 changes with temperature weakly, then at $\tau_c > T_1$ the rate constant must practically be independent of temperature.

Usually the interpretation of viscosity and temperature dependences of K_e in terms of strong or weak exchange can be done reliably enough. However, a situation must be mentioned in which some problems may arise. If a paramagnetic particle forms strong complexes with the solvent molecules or, as may happen in the case of charged particles, if it is associated with a counterion, then changing the solvent or the temperature may produce a change in its closest environment. This may result in an additional change in K_e due to a

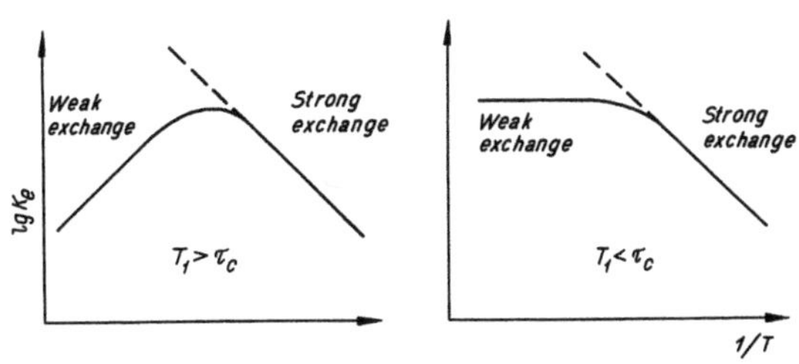

Fig.3.22. General view of the temperature dependence of the spin exchange rate constant for long and short relaxation times of the partner. T_1 is assumed to be independent of temperature

change in the exchange integral, in the steric factor, or in the diffusion mobility of the particle.

In conclusion it should be noted that the outlined criteria permitting one to separate the cases of strong and weak exchange have been repeatedly applied in analyzing experimental data. Numerous examples of such analysis will be considered in Sects.4.1,2.

4. Spin Exchange in Chemistry and Biology

Applications of spin exchange cover a number of problems of chemistry and molecular biology. Strong spin exchange is used in most of these applications. In this case spin exchange allows one to directly measure the number of binary collisions of paramagnetic particles in solution. A vast body of information has already been obtained on the numbers of binary collisions in simple systems that agree well with the modern theory of chemical kinetics in solutions. Collision rates have also been measured in complex systems, e.g., in solutions of polymers, in critical mixtures, etc., where no reliable theoretical calculations of these rates are possible as yet. An interesting field of application of spin exchange is the study of collisions of charged particles where modern theories are also not always applicable because of the insufficient knowledge of the detailed structure of interacting particles and of their close environment. In the field of molecular biology spin exchange supplies a simple way of establishing accessibility to collisions with other particles of paramagnetic centers present in biopolymers or specially introduced into a biological object as spin labels. Interesting applications are also associated with the study of the diffusion mobility of molecules in various biological structures, permeability of membranes, kinetics of some reactions, etc. Applications based on weak spin exchange are not yet so numerous. This field is, however, very attractive because it provides a unique opportunity of obtaining data on the exchange interaction during a collision and of thereby elucidating the intimate details of the collision act itself. Applications of spin exchange in exploring the collisions of atoms and radicals in gases, of triplet excitons in crystals, as well as of positroniums with paramagnetic particles are not included in this book. They can be, however, found in the original Russian edition of the book.

4.1 Study of Diffusion Collisions in Solutions

Diffusion collisions[1] of reacting particles are in most cases a necessary prerequisite for realizing chemical interaction between them in solutions. Therefore in the theory of liquid-phase reactions the rate constant of diffusion collisions K_D is one of the most important parameters. The comparison of the value of K_D with the experimentally determined rate constant K of a chemical reaction permits one to determine the number of collisions between reagents that are required for the chemical act to take place. For a reaction to occur at $K \ll K_D$ (kinetic region of a reaction) repeated collisions are needed, while at $K \simeq K_D$ (diffusion region of a reaction) each collision results in a reaction.

Hydrodynamic Model of Diffusion Collisions

A priori estimates of the diffusion rate constant are in most cases based on a hydrodynamic model in which the reagents are assumed to be spherical particles moving diffusionally in a viscous structureless liquid [4.1-3]. For such a model, according to the Smolukhovsky formula

$$K_D = 4\pi RD \ , \tag{4.1}$$

where $R = R_A + R_B$ is the sum of the Van der Waals radii of particles A and B; D is the sum of their diffusion coefficients. Within the framework of this model the diffusion coefficients $D_A(D_B)$ are described by the Stokes-Einstein equation

$$D_A = kT/6\pi\eta R_A \ , \tag{4.2}$$

where η is the solvent viscosity, T the temperature, and k the Boltzmann constant.

Subsituting (4.2) into (4.1) for the particular case of particles of the same size leads to the simple formula

$$K_D = 8kT/3\eta \tag{4.3}$$

or, expressing K_D in units l/mol·s ,

$$K_D = 8RT/3000\eta \ , \tag{4.4}$$

---

1 The term "collision" in liquid-phase kinetics has two meanings: 1) diffusion encounters and 2) collision of reagents in the solvent cage. Here the term will be used exclusively in the first sense.

where $R = 8.31 \cdot 10^7$ erg/deg. is the gas constant. A remarkable feature of (4.4) is the absence of any dependence on the size of reacting particles. This dependence turns out to be rather small for particles of different radii as well. Thus, at $R_A/R_B = 4$ the deviation of the value of K_D calculated according to (4.4) from that calculated according to (4.1) and (4.2) with allowance for the difference of the particle radii is no more than 50%. Therefore for estimations even at $R_A \neq R_B$ use can be made of (4.4) which for a given solvent at a given temperature yields the same value of K_D for all the molecules. For example, for water at room temperature (289 K) $K_D = 7.39 \cdot 10^9$ l/mol·s.

According to (4.3,4), to calculate K_D it is sufficient to know the viscosity of the solvent. In many cases, however, these equations obtained within the framework of a simplified model cannot claim to have an acceptable accuracy since they disregard the details of the structure of molecules and of their mobility in a liquid. Indeed, in many systems the mobility of small molecules is not determined by macroscopic viscosity. In solutions of polymers, for example, the viscosity can be associated with comparatively slow translational movement of huge macromolecules. But the mobility of small molecules in these solutions can be determined by much faster Brownian motion of small fragments of a polymeric chain. A concept of microscopic viscosity has, therefore, to be introduced. Besides, for many molecules a sphere is far from being a good approximation of their shape, the hydrodynamic radius in the Stokes-Einstein equation, in the general case, may be different from the Van der Waals one, etc.

Many difficulties connected with the application of the Stokes-Einstein equation (4.2) can be avoided if K_D is estimated from (4.1) by using experimental values of the diffusion coefficient. But even in this case uncertainties remain due to lack of knowledge about the true picture of molecular collisions in a liquid. Of primary importance are, therefore, the experimental methods of determing the constant K_D.

Potentialities of Spin Exchange

By now rather numerous reactions are known for which the rate constants approach the diffusion ones [4.4]. Examples of such processes can be found among reactions of hydrated electrons, proton transfer reactions, recombination reactions between atoms or free radicals, etc. However, systematic inquiry into the kinetics of superfast reactions in solutions is still a difficult experimental problem requiring complicated experimental methods for each particular class of reactions. Information on diffusion rate constants can also be obtained from the kinetics of other bimolecular processes whose rates may be controlled by diffusion, such as quenching of excited singlet and triplet molecules [4.4] and quenching of positronium [4.5]. Spin exchange also belongs to this kind of model processes. With the help of spin exchange one can study diffusion collisions only between paramagnetic particles. However, simplicity and high accuracy of the method based on spin exchange make it a very attractive instrument for solving many problems connected with diffusion collisions in solutions.

As follows from the discussion in Sect. 2.5, in the case of strong exchange the rate constant K_e of spin exchange is given by the equation

$$K_e = p_{max} \cdot fK_D , \qquad (4.5)$$

where f is a steric factor and p_{max} is the maximum value of the spin exchange efficiency corresponding to the extreme case of strong exchange. The values of p_{max} can be calculated from the equations given in Sect.2.5. From the analysis made there it follows that the value of p_{max} depends on the values of the electron spins of collision partners and on the broadening agent's relaxation time. If for the broadening agent $T_1 < \tau_c$, then regardless of the spin values $p_{max} = 1$. For the opposite case $T_1 > \tau_c$ the efficiency p_{max} can be found from (2.197). The numerical values of p_{max} for various practically interesting situations are summarized in Table 4.1.

Table 4.1. Theoretical values of the maximum spin exchange efficiency, p_{max}

Spin of indicator	Spin of broadening agent	p_{max}		Spin of indicator	Spin of broadening agent	p_{max}	
		$T_1 > \tau_c$	$T_1 < \tau_c$			$T_1 > \tau_c$	$T_1 < \tau_c$
1/2	1/2	1/2	1	1/2	1/2	1/2	1
1/2	1	16/27	1	1	1/2	2/9	1
1/2	3/2	5/8	1	3/2	1/2	1/8	1
1/2	2	16/25	1	2	1/2	2/25	1
1/2	5/2	35/54	1	5/2	1/2	1/18	1

An interesting theoretical prediction is the conclusion that for the interaction of particles with different spins the exchange broadening of the ESR spectra of the partners and hence the spin exchange rate constants calculated from these spectra will be different. It is important to have this conclusion in view when using (4.5) for calculating the constants of diffusion collisions. To test this conclusion experimentally, spin exchange was studied [4.6] between the radical 2,2,6,6-tetramethyl-4-oxopiperidine-1-oxyl (TANONE)

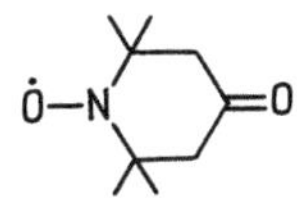

whose spin S = 1/2 and the $Mn(H_2O)_6^{2+}$ complex (S = 5/2). In one series of experiments the indicator was the free radical TANONE and the broadening agent was the manganese complex while in the other series, vice versa, $Mn(H_2O)_6^{2+}$ served as the indicator and TANONE as the broadening agent (Fig.4.1). The theory suggests that for such a pair of particles the ratio of spin exchange rate constants measured from different indicators must be $K_e(R)/K_e(Mn) = 11.7$ (the parentheses enclose the indicator). The experiments yielded the following results: $K_e(R) = (2.1 \pm 0.2) \cdot 10^9$ and $K_e(Mn) = (2.3 \pm 0.4) \cdot 10^8$ 1/mol · s and thus $K_e(R)/K_e(Mn) = 9 \pm 3$ - a value coinciding, within the limits of accuracy, with the above theoretical prediction.

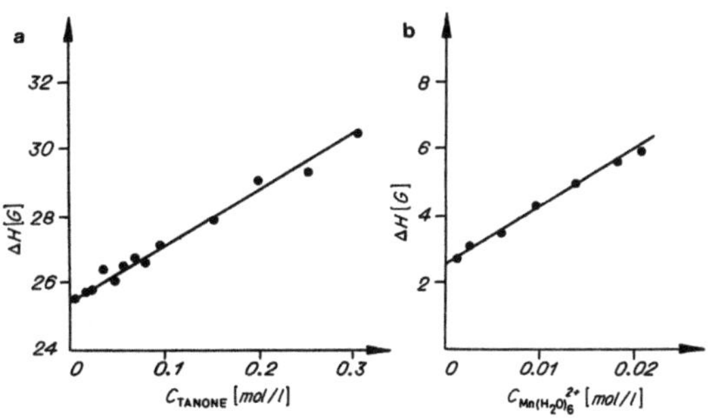

Fig. 4.1a,b. The dependence of the width of the low-field component of $Mn(H_2O)_6^{2+}$ ESR spectrum in aqueous solution on the concentration of TANONE radical (a) and the dependence of the width of the central TANONE component on the concentration of $Mn(H_2O)_6^{2+}$ (b) [4.6]

The possibility of theoretically calculating the values of p_{max} allows one, by using (4.5), to find the values of fK_D from experimentally measured spin exchange rate constants. In an extreme situation, when there are reasons to believe that no steric hindrance is present and that $f \cong 1$, experiments on spin exchange make it possible to find directly the values of the rate constant of diffusion collisions K_D. On the other hand, if a spin exchange is complicated by steric hindrance a possibility of estimating the value of the steric factor appears. In order to do so, it is naturally necessary either to determine the constant K_D from an independent experiment or to estimate it theoretically.

4.1.1 Collisions in Single-Component Solvents

To find out in practice the possibilities and limits of spin exchange applicability, let us first of all turn to spin exchange processes in the simplest systems, single-component solvents. In so doing one should choose, firstly, such processes for which the strong exchange condition is fulfilled, and, secondly, such solvents whose molecules are smaller in size than paramagnetic particles. The first condition indicates that the spin exchange rate is controlled by diffusion collisions; the second serves to create the most favorable conditions for the validity of the Stokes-Einstein equation (4.2) for calculating diffusion coefficients and, hence, for the validity of (4.3) or (4.4) for calculating K_D. This allows one to control the results of spin exchange measurements by comparing them with the results of theoretical calculations.

Numerous data on spin exchange rates in single-component solvents have been obtained, but they are dispersed in various publications which very often were not aimed directly at studying binary collisions. Thus, these data need additional analysis. First of all, it is necessary to find out which of the available experimental data refer to strong exchange. As we shall discuss in more detail later, as a rule, spin exchange between free radicals fits the case of strong exchange. The latter is also often observed for spin exchange between free radicals and metal complexes if the ligands do not produce too strong a screening effect. Thus, both systems may serve as a source of information on binary collisions in solutions.

Spin Exchange Between Radicals

Among the free radicals those of the nitroxide type have been studied most extensively. The fulfillment of the conditions of strong exchange for these radicals has been repeatedly verified experimentally. A detailed study was made of the viscosity dependence of the spin exchange rate between d-tert-butyl nitroxyl radicals (DTBN) $\dot{O}-N\begin{smallmatrix}\diagup C(CH_3)_3\\ \diagdown C(CH_3)_3\end{smallmatrix}$ in methylcyclohexane, n-pentane, and propane solutions [4.7]. Viscosity η was varied by changing external pressure and temperature. The conditions of strong exchange were found to be satisfied at $\eta \geq 0.3$ cP. The temperature changes of K_e have also confirmed the case of strong exchange for aqueous solutions of DTBN [4.10] (Fig. 4.2). In accord with the discussion in Sect.3.3 the linear character of dependence between K_e and T/η means that the conditions of strong exchange are indeed satisfied.

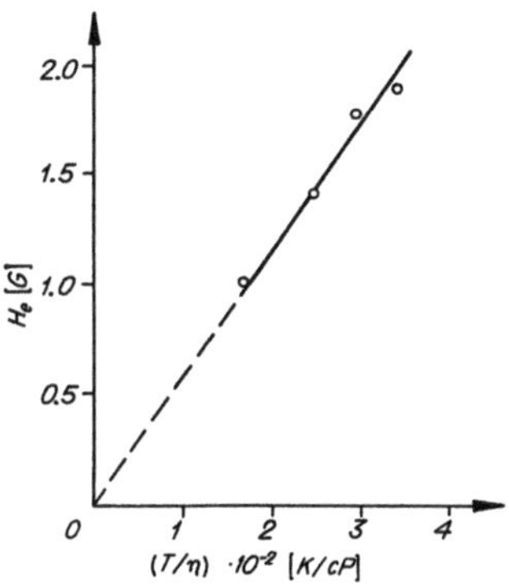

Fig. 4.2. The dependence of exchange broadening on T/η for an aqueous solution of DTBN ($2.4 \cdot 10^{-2}$ mol/l); 0.05 mol/l of K_2CO_3 is added to the solution [4.10]

Experimental data also show the strong exchange case for nitroxide radicals of a more complicated structure. The temperature dependence of the spin exchange rate constant has been studied for three nitroxide radicals as well as for a di-tert-butyl phenoxyl radical in methanol and ethyl-benzene [4.11]. The activation energies of K_e turned out to be close to those for the viscosity of solvents which corresponds to strong exchange. The same conclusion follows from the data on temperature dependence of spin exchange rate in solutions of various nitroxide radicals in malonic ester, decaline [4.12], and other organic solvents [4.13] as well as in water-glycerole mixtures [4.14]. On the strength of all this evidence it can be reliably concluded that spin exchange between nitroxide free radicals is usually strong. Some examples of spin exchange rate constants for nitroxide radicals are listed in Tables 4.2-4. Using these data, let us discuss how well the collisions of nitroxide radicals are described by (4.4). It should be borne in mind that for strong spin exchange between identical particles with $S = 1/2$ the efficiency of spin exchange $p_{max} = 1/2$. Therefore in the absence of steric hindrance the equation $K_e = K_D/2$ must be fulfilled. Comparison (see Table 4.2) shows that for the spin exchange between DTBN radicals this relation is indeed approximately fulfilled.

A similar conclusion may be drawn from the data for other nitroxide radicals. Table 4.3 lists values of $K_e \cdot \eta$, i.e., rate constants normalized to

Table 4.2. Comparison of observed rate constants of spin exchange K_e between DTBN radicals with the value of $K_D/2$ calculated by means of (4.4)

Solvent	t [°C]	$K_e \cdot 10^{-9}$ [l/mol · s]	$(K_D/2) \times 10^{-9}$ [l/mol · s]	References
Dimethoxy ethane	15	5.5	6.4	[4.9]
Tetrahydrofuran	15	5.0	6.3	[4.9]
Dimethyl Formamide	Room temp.	3.7	4.2	[4.15]
Water	24	2.4	3.6	[4.10]
Isooctane	30	6.3	7.6	[4.16]
Pentane	25	15	14.5	[4.8,9]

Table 4.3. Values of K_e in different solvents

Solvent	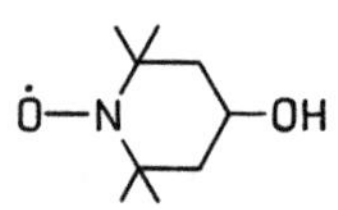		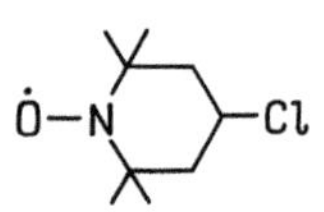	
	t [°C]	$K_e \cdot \eta \cdot 10^{-9}$ [cP · l/mol · s]	t [°C]	$K_e \cdot \eta \cdot 10^{-9}$ [cP · l/mol · s]
CCl_4	20-25	2.7	25	2.3
Dioxan	25	2.7	25	2.7
Ethanol	-32-70	2.4	25	2.6
Heptane	0-70	2.6		
Benzene	0-60	2.3		
Toluene	20	2.6		
Water	20-28	1.7		

Note: The listed values of K_e differ from the original ones by the factor $1/(1-\varphi) = 3/2$.

Table 4.4. Spin exchange rate constant for radicals of various structures in aqueous solution at room temperature [4.17]

Radical	$K_e \cdot 10^{-9}$ [l/mol · s]
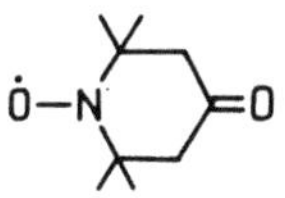	2.4
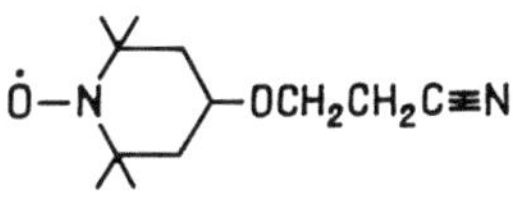	2.0
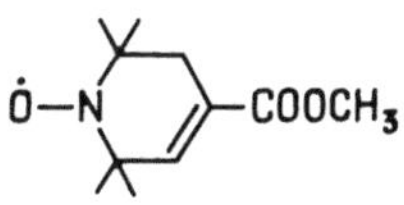	1.95
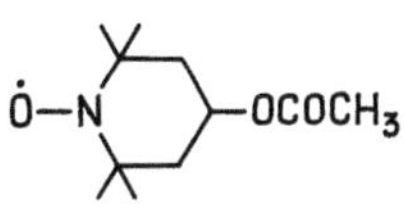	1.95
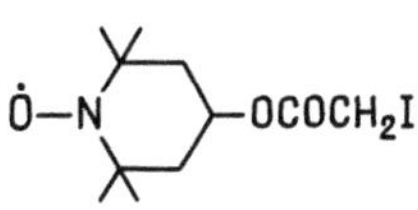	1.85

Note: The listed constants K_e differ from the original values by the factor $1/(1-\varphi) = 3/2$.

viscosity $\eta = 1$ cP for two radicals in various solvents. These values are close to the theoretical value of $K_D \cdot \eta/2 = 3.3 \cdot 10^9$ cP l/mol·s, calculated with the help of (4.4) for room temperature (298 K). For various radicals in the same (aqueous) solutions (see Table 4.4) the experimental spin exchange rate constants are also close to the value of $K_D/2 = 3.7 \cdot 10^9$ l/mol·s calculated according to (4.4).

The rate constants were also measured [4.18] for the spin exchange between the 2,2,6,6-tetramethyl piperidine-1-oxyl radicals in a series of solvents (Fig.4.3). In some of these solvents, as distinct from the systems listed in Table 4.3, the values of $K_e\eta$ are seen to deviate notably from the value obtained with the help of (4.4); however the differences between the experimental and calculated values of $K_e \cdot \eta$ are seen to be no more than twofold, most of the experimental points grouping around the calculated value (see Fig.4.3, straight line A). It was suggested [4.18] that the change in the value of $K_e\eta$ may be attributed to the transition into the region of weak exchange for solvents with a viscosity $\lesssim 1$ cP. This suggestion does not, however, agree with other data considered above. Most likely, the observed dependence of $K_e\eta$ on the nature of the solvent is due to other reasons, i.e., to the change of the steric factor f under the influence of solvation effects.

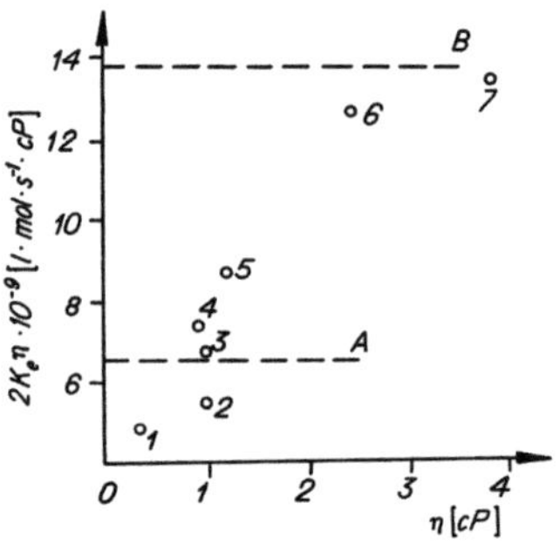

Fig. 4.3. The influence of viscosity on the value of $2K_e\eta$ for the radical 2,2,6,6-tetramethyl piperidine-1-oxyl in various solvents [4.18]. 1 - acetone; 2 - water; 3 - pyridine; 4 - decane; 5 - ethanol; 6 - dimethyl sulfoxide; 7 - i-butanol; A - the theoretical value of $K_D \cdot \eta$ according to (4.4); B - the value of $K_D \cdot \eta$ evaluated from the data on the self-diffusion of a diamagnetic analog

On the whole, the data available allow one to conclude that the results on the spin exchange between nitroxide radicals agree to an accuracy of a factor ~2 with the calculations of the numbers of diffusion collisions made according to (4.4) which is based on the Stokes-Einstein equation. Thus the example of nitroxide radicals demonstrates that spin exchange indeed can be used to determine, with a sufficient accuracy, the numbers of binary collisions in solutions.

The discrepancy observed for some systems between the rate constants of diffusion collisions calculated theoretically according to (4.4) and those obtained from spin exchange may be accounted for by two causes. On the one hand, it can be due to the inaccuracy of the hydrodynamic theory of diffusion collisions. The degree of accuracy of this simplified theory as applied to nitroxide radicals or molecules of similar structure can be judged from the following data. The diffusion coefficients D for nitroxide radicals I and II in heptane and decalin as measured by means of the capillary method were found [4.12] to exceed those calculated according to the Stokes-Einstein equation, the maximum discrepancy amounting to about two- or threefold.

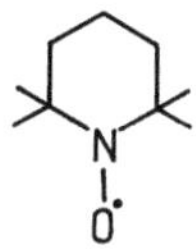

I

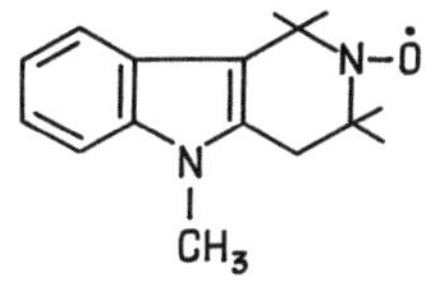

II

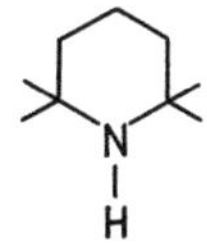

III

A similar discrepancy was found [4.18] for the diffusion coefficient of molecule III (diamagnetic analog of radical I) in various solvents with the aid of the nuclear spin echo method (see straight lines A and B in Fig.4.3).

The other reason for the discrepancy between the experimental data and the calculation may be the influence on the spin exchange of the steric factor whose value may be determined both by the structure of the radical itself and by the above-mentioned effects of solvating the NO group. The role of solvating effects can be estimated by comparing the spin exchange rate constants of a radical in different solvents. For example, the value of $K_e\eta$ for a water solution is less than for other solvents (see Table 4.3). This may be connected with the formation of a complex between a water molecule and the NO group of a radical [4.19] which hinders spin exchange. The influence of the radical structure is seen from the data on spin exchange between different radicals in the same solvent: introduction of bulky substituents into the piperidine ring diminishes the spin exchange rate constants (see Table 4.4). It may be suggested that collisions between these constituents do not lead to any spin exchange which decreases the steric factor.

Although the examples considered show that the influence of the steric factor in the case of nitroxide radicals must not be neglected, its value does not differ from $f = 1$ by more than an order of magnitude. This follows from a relatively small discrepancy between the values of diffusional collision constants obtained from spin exchange and those estimated theoretically. Most likely the absence of considerable steric hindrance is due to the fact that the strong exchange can be ensured not only by the direct contacts of the NO groups of two colliding radicals, but also by those of the NO group of one radical with the alkyl groups of the other (for more details see Sect. 4.2).

Information on the frequency of diffusion collisions in solutions can also be obtained by analyzing the data on spin exchange between anion radicals. Although anion radicals themselves are charged, in organic solvents they are commonly associated with counterions and therefore these are electroneutral particles which take part in collisions. A detailed analysis of this problem is made in Sect. 4.3; here we shall mention only that whenever the temperature dependence of the spin exchange rate was measured it indicated the situation of strong exchange to be realized. The absolute values of the spin exchange rate constants, just as for the nitroxide radicals, are close to those calculated according to (4.4).

Note that in the literature some systems are described for which rate constants of spin exchange between radicals are small. Thus the spin exchange rate constants of some nitroxide radicals in cumene were found to be by an order of magnitude smaller than usual [4.12]. A similar observation was made in studying the spin exchange of 2,4,6-tri-tert-butyl phenoxyl by the electron spin echo method [4.20]. The steric factor for spin exchange in toluene proved to be by about an order of magnitude smaller than in isopropanol where it is close to unity. In both solvents, judging by measurements at different temperatures, the conditions of strong exchange are fulfilled. A decrease was observed in the rate constant of spin exchange between DTBN radicals in an aqueous solution by about an order of magnitude after adding 0.4 M of electrolyte K_2CO_3 [4.10]. Finally, in the pioneering work [4.21] a conclusion is drawn that a spin exchange of DPPH in toluene is realized approximately by one out of ten collisions. At present it is not yet clear what is the reason for such a considerable decrease in the spin exchange rate constants. Note that a detailed investigation into this problem may prove useful, e.g., for establishing the factors affecting the rate of radical recombination in solutions.

Spin Exchange Between Free Radicals and Paramagnetic Complexes

Although the study of spin exchange in such systems was started later than that of exchange between free radicals, a considerable amount of experimental data has been obtained in this field by now. The attractive feature of the radical-complex ion systems is an additional and simple possibility to vary the structure of one of the partners (complex ion) by changing the composition of its first coordination sphere.

Table 4.5 lists the results of research into the spin exchange between chlorine-containing complexes of cobalt and free radicals - 2,2,6,6-tetramethyl-4-oxopiperidine-1-oxyl (TANONE), αα-dipenyl-β-picryl-hydrazyl (DPPH), and the cation of bisdiphenylchromium (BDC). The data show that for these systems a strong exchange occurs. Such a conclusion is made, e.g., for the spin exchange between the $Co(C_3H_7OH)_2Cl_2$ complex and TANONE, DPPH, and BDC on the basis of the temperature dependences of K_e [4.23]. For the spin exchange between the complexes $Co(ROH)_2Cl_2$ and DPPH in various alcohols a nearly

Table 4.5. Rate constants of spin exchange between free radicals[a] and chlorine-containing cobalt complexes (t = 20 °C) [4.22-24]

No. of system	Complex	Solvent	$K_e \cdot 10^{-9}$ [l/mol · s]			$K_D \cdot 10^{-9}$ [l/mol · s]
			TANONE	BDC	DPPH	
1	$CoPy_2Cl_2$	CH_3NO_2	2.5	7.7	-	9.7
2	$Co(\alpha mPy)_2Cl_2$	CH_3NO_2	-	9.1	-	9.7
3	$Co(\alpha mPy)_2Cl_2$	αmPy [b]	-	5.0	-	6.6
4	$CoPy_4Cl_2$	Py	1.4	2.4	-	6.6
5	$Co(C_2H_5OH)_2Cl_2$	C_2H_5OH	-	-	3.0	5.3
6	$Co(C_3H_7OH)_2Cl_2$	C_3H_7OH	1.0	2.0	2.0	2.9
7	$Co(C_4H_9OH)_2Cl_2$	C_4H_9OH	-	-	1.5	2.2
8	$Co(iso\text{-}C_4H_9OH)_2Cl_2$	iso-C_4H_9OH	-	-	1.5	1.6
9	$Co(C_8H_{17}OH)_2Cl_2$	$C_8H_{17}OH$	0.45	-	0.1	0.7

a

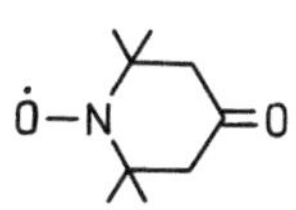

TANONE

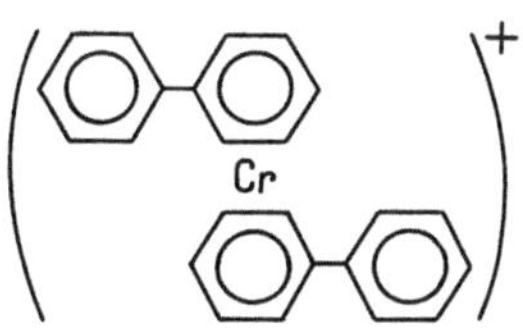

BDC

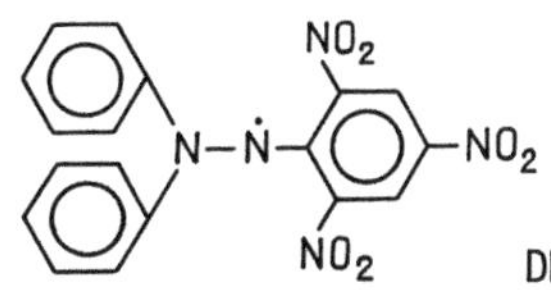

DPPH

b αmPy is α-methylpyridine; the viscosity of αmPy is taken to be equal to that of pyridine.

linear dependence $K_e \sim \eta^{-1}$ was found [4.22], which also corresponds to the strong exchange (Fig.4.4). In this case the strong exchange seems to be provided by the presence of a chlorine ion in the first coordination sphere of cobalt. The high efficiency of chlorine is naturally accounted for by the high spin density $\rho \cong 0.05$ [4.25] on this ion which considerably exceeds the spin densities on the protons of organic ligands (10^{-2}-10^{-4} [4.26]). It is also essential that the chlorine ion is relatively large and, hence, is readily attainable in colliding with free radicals.

As follows from Table 4.5, in a number of cases the experimentally obtained rate constants of spin exchange approach the values of the constants K_D calculated according to (4.4). (Note that the maximum value of the spin

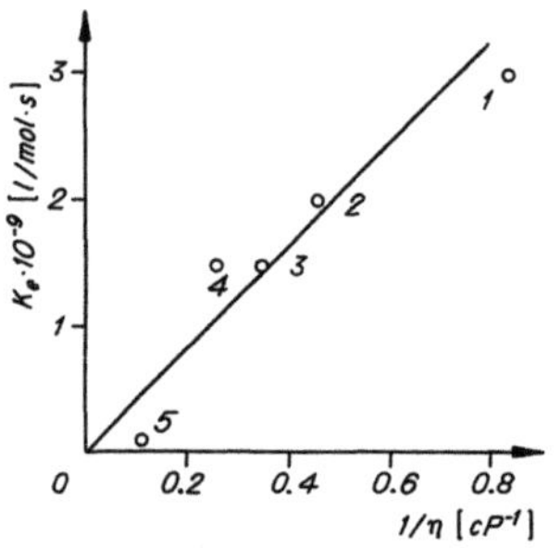

Fig. 4.4. The dependence of the rate constant of spin exchange between DPPH and $CoCl_2(ROH)_2$ on the reciprocal viscosity of the solvent [4.22]. 1 - ethanol; 2 - propanol; 3 - butanol; 4 - isobutanol; 5 - octanol

exchange efficiency p_{max} is in this case equal to one, since the cobalt complexes have a short relaxation time). At the same time there are cases where the measured constants are several times smaller than those calculated. As a rule, this difference may be attributed to the appearance of a steric factor. Thus, the rate constants of the spin exchange involving TANONE generally have a smaller value than that for the exchange between the same complexes and DPPH or BDC. This can be explained by the different character of spin density delocalization for these radicals. For nitroxide radicals where the spin density is but insignificantly delocalized from the NO group to the other atoms the strong exchange with the complex appears to be provided for by such collisions only when a chlorine ion and an NO group chance to be in contact. On the contrary, in DPPH and BDC the spin density is delocalized over many atoms of aromatic rings which provides better steric possibilities for contacts with chlorine ions.

Another example of the influence of steric restrictions may be the value of the rate constant for the spin exchange of DPPH with $Co(C_8H_{17}OH)_2Cl_2$ which is small compared with K_D. The steric restrictions here seem to be due to bulky alkyl groups of octanol. Finally, the steric restrictions appear to be responsible also for the difference of the rate constants of the spin exchange of the $CoPy_2Cl_2$ and $CoPy_4Cl_2$ complexes with BDC (cf systems 1 and 4, Table 4.5). In the tetrahedral complex the chlorine ions occupy a larger part of the surface and are more attainable than in the octahedral one.

The attention of many investigators is attracted by the spin exchange of aquo-complexes of the iron group ions with nitroxide radicals. Recently, the flow technique has also been used to study the spin exchange of aquo complexes with unstable radicals. Some of the results obtained for nitroxide radicals [4.27-29] are summarized in Table 4.6. Similar results have been obtained in [4.14,30]. An interesting result is a considerable difference of the spin exchange rate constants for various ions. Note also that the values of the constants K_e are appreciably smaller than K_D. This difference becomes especially marked if one takes into consideration [4.28] that the experimental coefficients of diffusion for aquo ions are about twice as large as the ones calculated according to the Stokes-Einstein equation.

The spin exchange rate constants of aquo ions with three unstable radicals are listed in Table 4.7 [4.28]. These constants were determined by the stationary saturation technique from the influence of the ions on the radical relaxation times T_2 [4.31]. In this case the qualitative features of the spin exchange remain the same as for the nitroxide radicals. The absolute values of the constants, however, are about twice as large which may be due to the greater mobility and smaller steric restrictions for the active radicals studied.

The problem of meeting the conditions of strong exchange for processes involving aquo complexes has been studied only for the Co^{2+} and VO^{2+} ions. Study was made of the temperature behavior of the spin exchange rate constant in the system $Co(H_2O)_6^{2+}$ + TANONE and the product $K_e \cdot (\eta/T)$ was shown to be independent of temperature [4.23]. This points to the strong exchange. The low absolute value of K_e appears here to be due to steric reasons. It has been suggested that because of the short T_1 of the ion the contacts of the radical with the hydrogen atoms of the coordinated water molecules stop contributing to the spin exchange. What remains are the contacts with the oxygen atoms of the water molecules which are likely to be sterically hindered. In studying the interactions of the vanadyl sulfate $VO(H_2O)_5SO_4$ with the free radical 2,2,6,6-tetramethyl-4-oxypiperidine-1-oxyl a deviation was found from the regularities of strong spin exchange [4.29]. The spin exchange rate constant in this system increases with increase in temperature from 0 to 25 °C. However in this case the value of $K_e \cdot (\eta/T)$ falls rather than remains unchanged as should have been the case for strong exchange. Indirect evidence in favor of weak exchange in the case of the Ti^{3+} ion may be the low absolute value of the spin exchange rate constant for this ion.

Thus, the deviations of the rate constants of spin exchange from the diffusional rate constants, observed for the collisions between the aquo complexes of the iron group ions and the nitroxide radicals, may be due to both the steric hindrance and the transition into the region of weak exchange. For different ions there appear to be different situations. Further studies will probably make it possible to reveal the intimate details of the mechanism of collisions of free radicals with aquo ions in water solutions. The significance of these data for an understanding of the mechanism of redox reactions between these particles hardly requires any comment.

Table 4.6. Rate constants of spin exchange between nitroxide radicals and aquo complexes of iron group ions

Ion	Radical	$K_e \cdot 10^{-9}$ [l/mol · s]	References
Cr^{3+}		1.1	
Mn^{2+}		2.0	
Fe^{2+}	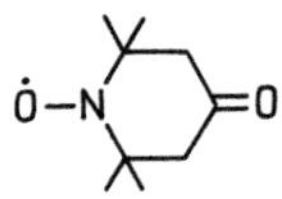	0.9	
Co^{2+}		0.9	[4.27]
Ni^{2+}		1.1	
Cu^{2+}		1.4	
Cr^{3+}		1.2	
Mn^{2+}		2.2	
Co^{2+}	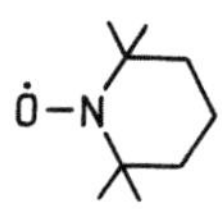	0.9	
Ni^{2+}		1.3	[4.28]
Cu^{2+}		1.6	
Ti^{3+}		0.3^a	
VO^{2+}	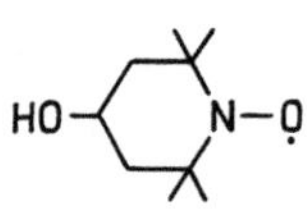	0.8	[4.29]

[a] Measured in a flow system.

Table 4.7. Rate constants of spin exchange between unstable radicals and aquo complexes of iron group ions ($K_e \cdot 10^{-9}$ [l/mol · s]) [4.28]

Ion	$CH_3\dot{C}HOH$	$\dot{C}H_2COOH$	$(CH_3)_2\dot{C}OH$
Mn^{2+}	3.2	2.7	2.5
Ni^{2+}	2.2	1.9	2.2
Co^{2+}	2.6	2.7	2.0
Ti^{3+}	0.9	0.7	0.5
Cr^{3+}	1.5	1.5	1.4

In the literature there is a vast body of experimental data on the rate constants of the spin exchange of nitroxide radicals with various complex compounds containing organic ligands in the first coordination sphere. The character of spin exchange in these cases was not studied, as a rule. For

Table 4.8. Rate constants of spin exchange of TANONE with some complexes at room temperature

Complex	$K_e \cdot 10^{-9}$ [l/mol · s]	References
$Cu(NH_2CH_2CH_2NH_2)_2^{2+}$	2.6	[4.14]
$[CuN(CH_2CH_2NH_2)_3OH]^+$	2.5	[4.32]
$Cu(NH_2CH_2CH_2N(C_2H_5)_2)_2^{2+}$	1.0	[4.32]
$Cu(NH_2CH_2COO)_2$	2.0	[4.30]
$Ni(NH_2CH_2CH_2NH_2)_2^{2+}$	2.1	[4.14]
Fe(III) hemin pyridinate	2.4	[4.30]
Fe(III) ethioporphyrin pyridinate	2.6	[4.30]
Cu(II) " "	2.7	[4.30]
VO(II) " "	1.5	[4.30]

many of these systems, however, the spin exchange rate constants are close to the diffusion constants of binary collisions and probably correspond to the case of strong exchange. The appropriate examples are summarized in Tables 4.8-10.

All the complexes listed in Table 4.8 have the electron spin $S = 1/2$ and a sufficiently long relaxation time, so that the exchange efficiency $P_{max} = 1/2$. A comparison of K_e with the calculated value $p_{max} \cdot K_D = 3.7 \cdot 10^9$ l/mol · s for aqueous solutions shows these quantities to be in many cases close to each other. In the original publications [4.30,32] one can find other examples of close-to-diffusional constants of spin exchange rates for both water and various organic solvents. The absence of the screening action of organic ligands in the systems under consideration may be due to several reasons. Firstly, the spin density on the hydrogen atoms of NH_2 groups of amines and in hemin and porphyrin aromatic rings may be sufficient to ensure a strong exchange. Secondly, most of the complexes under discussion have a flat structure. In such complexes the attack of a radical may also proceed along the axial positions displacing the labile molecule of the solvent.

Tables 4.9 and 4.10 list rate constants of the spin exchange of nitroxide radicals with a dibenzyl chromium cation and with iron (III) acetylacetonate [4.17]. Here possibly a strong exchange occurs as well, since the rate con-

Table 4.9. Rate constants of spin exchange between nitroxide radicals and $Cr(C_6H_6)_2^+$ in aqueous solution [4.17]

Radical	$K_e \cdot 10^{-9}$ [l/mol · s]
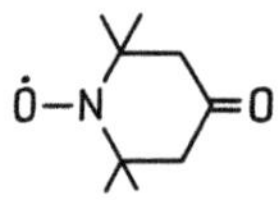	2.8
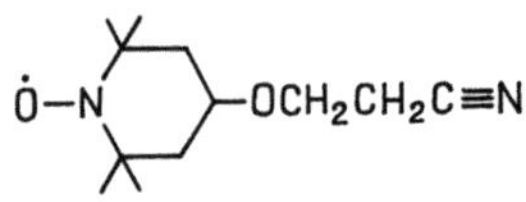	2.5
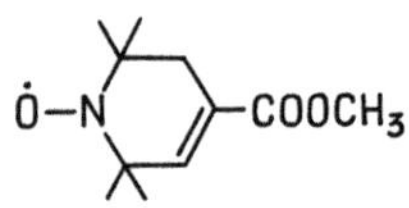	2.5
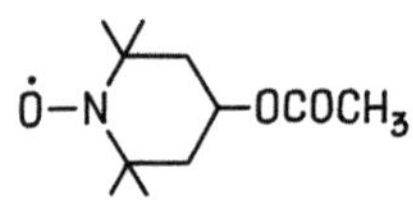	2.5
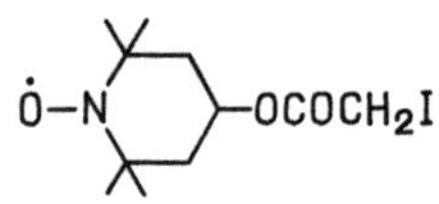	2.1

stants observed are close to the maximum values $K_e = p_{max} \cdot K_D$ corresponding to the case $f = 1$ [K_D was calculated according to (4.4)]. Indeed, in the first case [$K_D = 6.4 \cdot 10^9$ l/mol · s (20°); $p_{max} = 0.5$ as the spin of the broadening agent $S = 1/2$] we have $p_{max} \cdot K_D = 3.2 \cdot 10^9$ l/mol · s, and in the second case [$K_D = 5.4 \cdot 10^9$ l/mol · s (20°); $p_{max} = 0.65$ as $S = 5/2$] $p_{max} \cdot K_D = 3.5 \cdot 10^9$ l/mol · s. Note that in both series the rate constant decreases with increase in the size of the substituent just as is the case of spin exchange between radicals.

Thus, for radical-paramagnetic complex collisions the maximum values of rate constants of spin exchange are close to the theoretical values of $p_{max} \cdot K_D$ where K_D is calculated according to (4.4).

On the whole one can draw the following conclusion from the data discussed in the present subsection. On the example of the simplest systems - monocomponent liquids where the application of (4.1-4) is most justified - spin exchange has been shown to permit the determination with an adequate accuracy

Table 4.10. Rate constants of spin exchange between nitroxide radicals and Fe(III) acetylacetonate in ethanol [4.17]

Radical	$K_e \cdot 10^{-9}$ [l/mol · s]
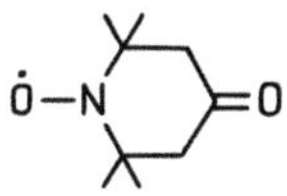	4.5
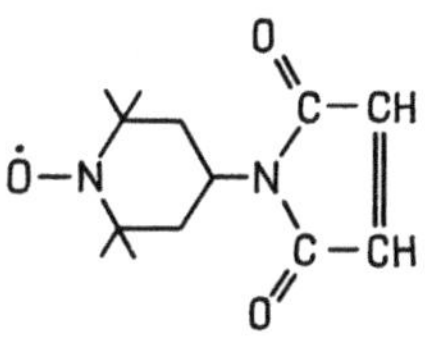	3.5
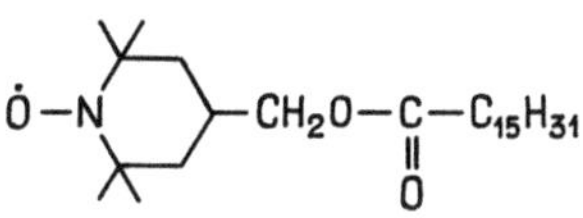	2.9
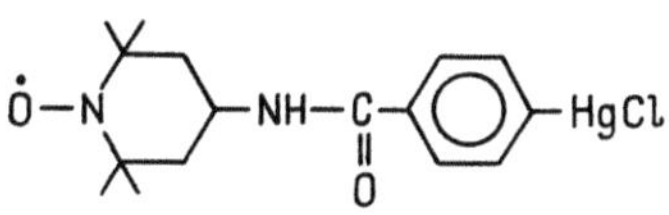	2.3

of the number of binary collisions. Thus it is possible to use spin exchange for measuring the number of collisions in more complex systems for which even rough theoretical calculations are so far impossible. To these belong, for example, solutions of particles whose size is smaller than that of the molecules of the solvent. In this case the application of the Stokes-Einstein equation meets with serious objections. As other examples one may mention polymer solutions, multicomponent liquids, biological membranes, etc. On the other hand, spin exchange turned out to be sensitive to such fine details of the collision mechanism as steric restrictions and solvation. Hence, there is a possibility to also study these features whose importance for chemical reactions in solutions is well known.

4.1.2 Collisions in Complex Systems

As one passes over to complex systems spin exchange turns from the subject of research as it was in simple situations into a unique source of information on diffusion collisions. Some complex systems that can be studied using spin exchange include binary solvents, liquids in critical states, polymers, and polymeric radicals in solutions. In later subsections collisons in solutions of electrolytes and in biological systems will also be considered. Paramagnetic particles in these investigations serve as probes enabling one to estimate the molecular mobility in a system.

Binary Solvents

Collisions of particles in binary solvents may be of far more complicated character than in one-component systems. The complicating factors include, among others, selective solvation. The role of this factor may be especially important if the colliding particles in question possess quite different polar properties or different ability for specific solvation. The situation may become even more complicated due to the possible microheterogeneous structure of the binary solvents themselves [4.33], i.e., they may consist of microregions enriched in one or the other component. The part played by such factors in the kinetics of diffusive collisions is far from known and here spin exchange opens apparent perspectives.

The data on spin exchange studies of diffusion collisions in binary solvents are not numerous, but rather important [4.13,29]. Interesting results were obtained from comparing the rates of exchange for spin probes of essentially different solubilities in the components of the binary solvent [4.13]. The spin probes in these experiments were the 2,2,6,6-tetramethyl-4-oxypiperidine-1-oxyl (R) and 2,2,6,6-tetramethyl-4-chloropiperidine-1-oxyl (R') radicals

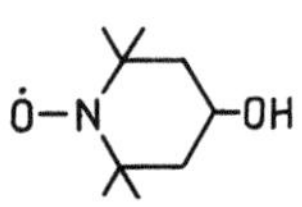

(R)

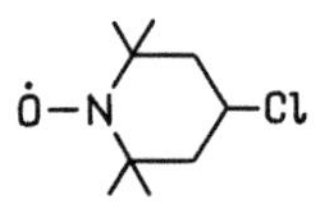

(R') .

The results of the experiment are presented in Fig.4.5. The multiplication of the spin exchange constant K_e by the viscosity η excludes the trivial dependence of the number of binary collisions on the composition resulting from the viscosity changes.

The experimental dependences obtained may be divided into two groups: smooth curves I and II and curves III and IV with extrema. The existence of the extrema can be explained [4.13] within the framework of the hypothesis of microheterogeneity in binary systems. Assume, e.g., that in the water-dioxan system there are water globules separated by the dioxan phase. Since the radical R' is poorly soluble in water it must be concentrated in the dioxan phase thus leading to an increase in the number of binary collisions

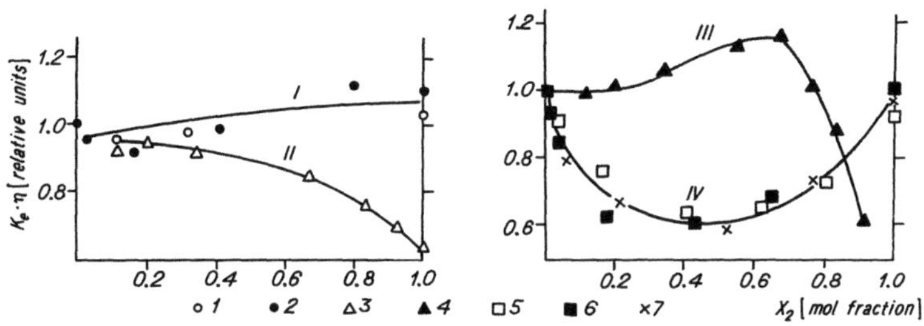

Fig. 4.5. The dependence of $K_e \cdot \eta$ on the mole fraction X_2 of the second component in binary solutions [4.13]. 1 - CCl_4/dioxan - R radical; 2 - CCl_4/ethanol - R' radical; 3 - dioxan/H_2O - R radical; 4 - dioxan/H_2O - R' radical; 5 - CCl_4/ethanol - R radical; 6 - toluene/ethanol - R radical; 7 - heptane-ethanol - R radical

and hence in K_e . In the ethanol-nonpolar solvent systems the radical R, due to its better solubility in ethanol, is accumulated in the alcohol globule which isolates the radicals from one another and decreases the spin exchange rate constant.

Interesting information was also obtained by studying spin exchange between two paramagnetic probes which were simultaneously introduced into a water-dioxan mixture [4.29]. One of them, a free radical R', is readily soluble in dioxan and practically insoluble in water, while the other, $VOSO_4$ or $Cu(en)_2(NO_3)_2$, is more soluble in water than in dioxan. The results (Fig. 4.6) agree qualitatively with the hypothesis that the radicals and the complex ions are spatially separated. Indeed, when dioxan is added to water the spin exchange constants for R' + VO^{2+} and for R' + $Cu(en)_2^{2+}$ decrease, the scope of the change exceeding markedly the effect that could be produced by a change in viscosity. Note that for the system R' + $Cu(en)_2^{2+}$ the condition of strong exchange was determined by the fulfillment of the viscosity criterion. Therefore curve 3 in Fig.4.6 reflects directly the change in the number of binary collisions between R' and $Cu(en)_2^{2+}$ with the change in the composition of the water-dioxan mixture. The rate constants of the spin exchange between the same particles, VO^{2+} + VO^{2+} and R' + R', behave inversely: they increase as dioxan is added to water.

These examples agree well with the hypothesis of the heterogeneity of some binary solvents but do not prove this phenomenon. The data of Fig.4.6 may, e.g., be accounted for in the following way. It is probable that since the radical R' prefers a dioxan surrounding while the complex prefers a water surrounding, the two particles tend to avoid each other when diffusing in the mixed solvent. But whatever may be the details of interpretation, even

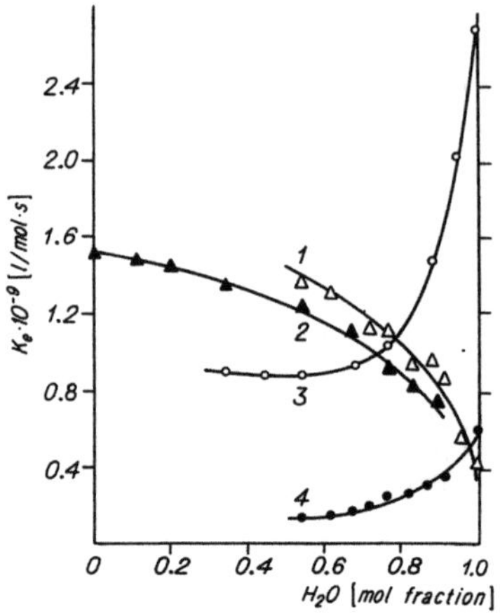

Fig. 4.6. Rate constants of spin exchange in water-dioxan mixtures for some systems [4.29]. 1 - $VOSO_4 + VOSO_4$; 2 - R' + R'; 3 - R' + $Cu(en)_2$; 4 - R' + $VOSO_4$

these, so far not numerous, experiments allow us to draw a conclusion of importance for chemical kinetics: the changes in the number of collisions in binary mixtures are not described by changes in viscosity. Thus, experiments on spin exchange make it possible to single out the simplest factor affecting the rate of a chemical reaction in mixed solvents: the number of binary collisions.

Liquids Near Critical Points

Anomalous properties of liquids near critical points attract the attention of many authors. Spin exchange between paramagnetic particles under such conditions can be used as a useful tool for elucidating the nature of these properties on a molecular level. It is of interest, e.g., to study the nature of the anomalies of the transport coefficients near the critical composition of the two solvents and at the liquid-gas critical point. If the anomalies are connected with the changing of molecular mobility then they must reveal themselves in the unusual behavior of the spin exchange rate constants.

Several studies by means of spin exchange of the mobility of molecules at the critical compositions were made [4.16,34]. For example, a mixture of heptane- and acrylonitrile (39.5 weight %) was studied [4.34] at a critical temperature of 18.0 °C. The paramagnetic probe was the nitroxide radical 2,2,6,6-tetramethylpiperidine-1-oxyl

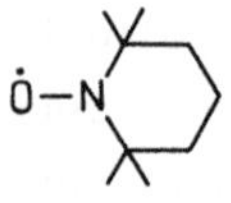

readily soluble in either component. The spin exchange rate constant measured with a step of 0.02° in the vicinity of the critical point did not reveal any peculiarities (Fig.4.7). Consequently, the translational mobility of the radical at the critical point does not undergo any sharp changes, and the anomalies registered by viscosimetric methods are not connected with the decrease in the molecular mobility.

A similar result was obtained for a mixture of isooctane (mole fraction is 0.58) and n-perfluoro-heptane [4.16] whose critical temperature is $\simeq 24\,^{\circ}C$. The paramagnetic probe was the DTBN stable radical. In these experiments no anomalies of the exchange broadening of the ESR lines near the ciritical temperature have been found either. At the same time it has been shown that if translational mobility were determined by macroscopic viscosity which increases at the critical point then the decrease in spin exchange rate would be observed to go far beyond the limits of experimental errors.

To explain the absence of anomalies in experiments on spin exchange modern theories were used [4.16] according to which in binary solutions the anomalous contribution to macroscopic viscosity at a critical point is due to large-scale fluctuations of concentrations. At a critical point the scale of these fluctuations is of the order of 1000 Å, and their lifetime is about 10^{-3} s. These fluctuations cannot manifest themselves in spin exchange [4.16] as the size of the paramagnetic probe is too small, and the collision frequencies ($\sim 10^8\ s^{-1}$) characteristic of the spin exchange experiments are too high.

The fluctuations of concentrations appear to reveal themselves in another phenomenon found for critical mixture: as distinct from pure components or noncritical compositions the temperature dependence of exchange broadening is expressed far more weakly than follows from the Stokes-Einstein equation [4.16]. Such behavior cannot be ascribed to a transition into the region of weak spin exchange. In a critical mixture the diffusion of the probe (DTBN) is assumed [4.16] to involve mostly the regions rich in isooctane in which the solubility of DTBN is higher than in the other component, perfluoroheptane. As temperature decreases this selective solubility increases which partially makes up for the decrease in the exchange broadening caused by an increase in viscosity.

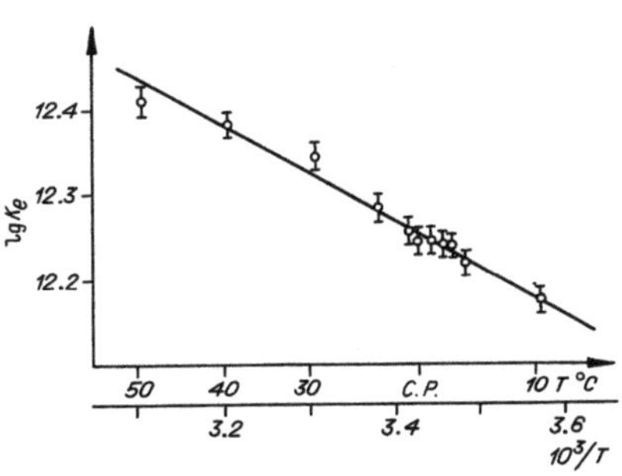

Fig. 4.7. The temperature dependence of the spin exchange rate constant K_e ($cm^3 \cdot s^{-1}$) of the radical in an acrylonitrile-heptane mixture of the critical composition [4.34]. There is nothing special with the behaviour of the rate constant in the critical region

The fluctuations of concentrations at the critical point might be revealed more distinctly in experiments on spin exchange between different probes each of which would have a preferential solubility in one of the mixture components. Large-scale fluctuations near the critical point would lead to a sharp drop of spin exchange rate due to the localization of each of the probes in regions rich in the favorable component. In its design such an experiment would be similar to the experiments on diffusion collisions in binary noncritical mixtures described in the previous subsection.

The molecular mobility at the critical liquid-gas point was also studied by the spin exchange method [4.35]. A small amount (0.2 weight %) of paramagnetic probe, 2,2,6,6-tetramethylpiperidine-1-oxyl was introduced into ethane or ethylene. The technique applied allowed the ESR spectrum to be recorded at various points of a long ampule which was carefully thermostated. To eliminate temperature gradients in the sample, an inside mixer was used.

In the vicinity of the critical point the spin exchange rate constant in the gas phase was found to exceed that in the liquid, the change in the rate constant in the transition region being of an irregular character. A detailed analysis showed these irregularities to be unrelated to the molecular mobility near the critical point. They appear to be due to gravitational effects: formation, at the critical point, of density fluctuation clusters and their precipitation to the bottom of the ampule. The cluster formation near the critical point has been most distinctly observed at a small temperature gradient ($\sim 0.1^{\circ}$) along the sample, and with no mixing [4.35]. Under such conditions the concentration of radicals in the clusters becomes essentially higher than in the intercluster medium. As a result, two ESR signals are simultaneously observed near the critical point: a triplet corresponding to the intercluster medium, and an exchange-broadened singlet corresponding to the clusters (Fig.4.8). Far from the critical point only a triplet signal is observed. On the basis of these results and of the analysis of the data available in the literature the conclusion was drawn [4.35] that at the critical point there must be two types of diffusion: diffusion of molecules inside the clusters undergoing no anomalous changes, and intercluster diffusion whose rate is changed in the vicinity of the critical point.

Thus, spin exchange opens up some interesting possibilities for research into molecular mobility and into spatial distribution of particles under critical conditions. They may be of interest for studying not only the nature of critical states but also the peculiarities of the kinetics of chemical reactions at critical points - a field that is only beginning to attract the attention of explorers.

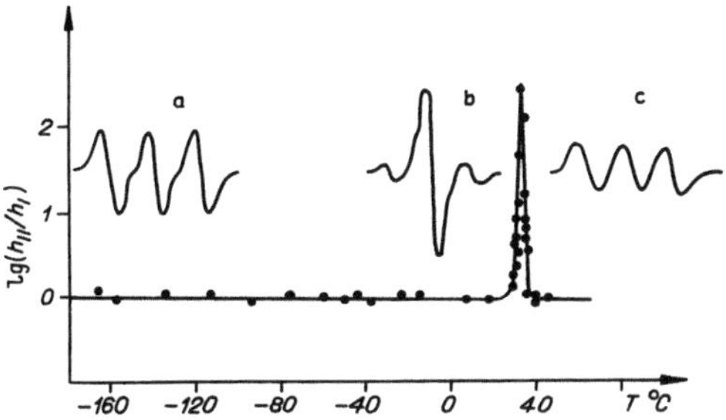

Fig. 4.8. The temperature dependence of the ratio of the intensities of the central (h_{II}) and the side (h_I) components of the ESR spectrum for a solution of nitroxide radical in ethane [4.35] (a-c is the change in the profile of the ESR spectrum in passing through the critical point)

Polymeric and Heterogeneous Systems

Characteristic of macromolecules is a set of various types of motion with various rates from a fast micro-Brownian motion of the short segments of a polymer chain up to a comparatively slow translation of macromolecules as a whole. The diffusion mobility of small molecules in macromolecular solvents is mainly determined by the micro-Brownian motion [4.36]. Therefore macroscopic viscosity in such systems, as a rule, has nothing to do with molecular mobility. The study of diffusion collisions of radicals with the help of spin exchange provides good illustrations of this statement and shows how spin exchange can be used to measure microscopic viscosity in polymeric systems.

Let us discuss the results available in the literature. When passing from chlorobenzene to a 20% solution of polystyrene in chlorobenzene, the macroscopic viscosity changes from 0.5 to 570 cP (at 60 °C), i.e., by more than three orders of magnitude. Under the same conditions the spin exchange rate constant for the nitroxide radical 2,2,6,6-tetramethyl-4-oxypiperidine-1-oxyl (and, hence, the microscopic viscosity as well) changes very little - from $3.7 \cdot 10^9$ to $2.4 \cdot 10^9$ l/mol · s [4.37]. For the same radical in another high molecular solvent-vaseline - the spin exchange rate constant is by one or two orders of magnitude higher than could be predicted from the value of the macroscopic viscosity [4.12]. A similar result was obtained in studying the 2,4,6-tri-tert-butyl phenoxyl radical in silicon oils by the electron spin echo method [4.20]. The experimentally obtained spin exchange rate constants for the two samples corresponded to the effective values of the microscopic viscosity 1.9 and 3.2 cP instead of the macroscopic values 10 and 100 cP, respectively.

To understand the mechanism of motion of small molecules in high molecular solvents it is of interest to consider a relation found between the translational and rotational diffusion [4.12]. The former was studied by the spin exchange between nitroxide radicals, and the latter by the broadening of hf components of these radicals caused by anisotropic interactions in the absence of spin exchange. Irrespective of the nature of the solvent the constant of the translational diffusion of a radical turned out to be proportional to that of its rotational diffusion. This relation is quite natural for systems obeying the hydrodynamic model of diffusion motion. However, the polymeric systems do not fit this theory since for them the diffusion constants are larger than the Stokes ones by several orders of magnitude. The observed correlation means that both the translational and rotational movements of radicals in polymeric systems proceed via one and the same mechanism and have equal activation energies [4.12]. The regularity in question may also be formulated as follows. At a given rotational mobility (this can be selected by temperature adjustment) the diffusion coefficients of radicals and, hence, the rate constants of their spin exchange must be the same (Fig.4.9). It is seen that for all polymers the concentration dependences of the broadening which is due to spin exchange are the same if the rotational (orientation) relaxation times of radicals are the same.

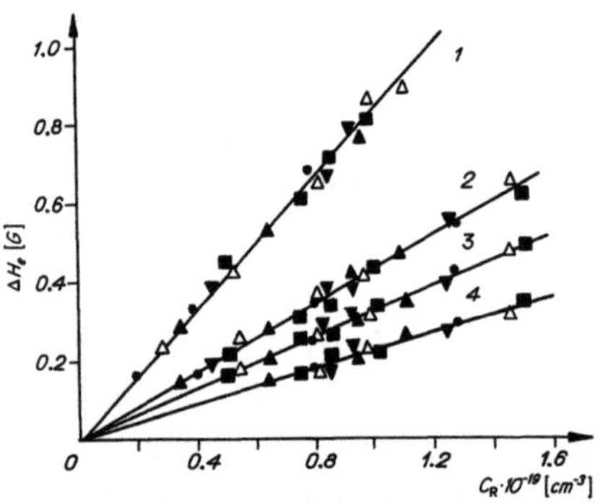

Fig. 4.9. The concentration dependences of the exchange broadening in the ESR spectrum of 2,2,6,6-tetramethyl piperidine-1-oxyl radical in some polymers at different rotational relaxation times of the radical [4.12]. The values of τ_{rot} [s] are: 1 - 10^{-10}; 2 - $2.1 \cdot 10^{-10}$; 3 - $3.2 \cdot 10^{-10}$; 4 - $4.2 \cdot 10^{-10}$

This correlation may find interesting applications in studying the distribution of radicals in heterogeneous systems [4.12]. In the case of the nonuniform distribution of radicals in the substance the concentration broadening of the ESR lines caused by spin exchange will be larger than is predicted by the dependence presented in Fig. 4.9. In isotactic, as opposed to atactic, polypropylene with a 60% crystallinity a higher value was obtained for the spin exchange rate constant (Fig.4.10). This indicates that the radicals are concentrated in the amorphous region of this polymer. This conclusion was confirmed by independent experiments on measuring the dipole-dipole broadening in frozen solutions at 77 K [4.12].

A similar technique has been applied [4.38] to study the structure of the system polymer (butylcaoutchouc) - low molecular solvent (toluene). For this purpose the behavior of a paramagnetic probe - the TANONE radical - was studied in this system at various ratios of the components in a wide range of temperatures. For ratios close to 1:1 at low temperatures the spin exchange rate constants became several times as large as the values that can be predicted by means of the curves in Fig.4.9 using the independently measured rotational mobility. Hence a conclusion was drawn that in the butylcaoutchouc-toluene mixture a decrease in temperature results in formation of microphases of two components, the free radical concentrating in the low molecular phase.

Thus, spin exchange opens an opportunity of studying the distribution of radicals in heterogeneous systems. These data are of great importance for the kinetics of chemical reactions in such systems since the selective concentration of reagents must influence markedly the rate of reactions. This influence is considered [4.39] to account for an increase in the rate of chemical reactions for a number of systems when they are frozen.

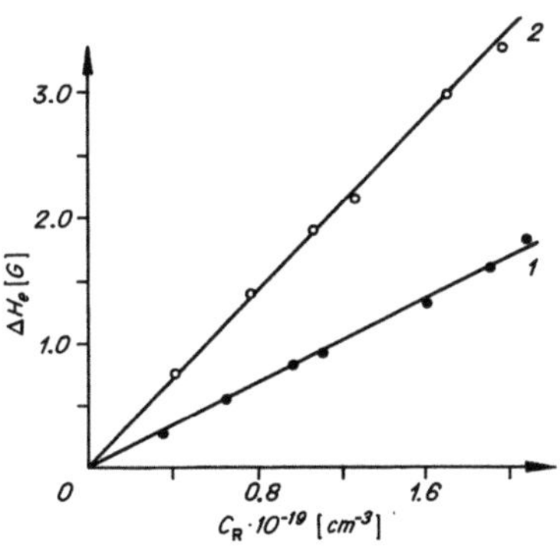

Fig. 4.10. The concentration dependences of the exchange broadening in the ESR spectrum of 2,2,6,6-tetramethyl piperidine-1-oxyl radical in atactic (1) and isotactic (2) polypropylene at $\tau_{rot} = 1.0 \cdot 10^{-10}$ s [4.12]

Collisions of Macroradicals

Spin exchange between macroradicals in solutions offers a convenient method of modeling such an important and insufficiently known process as recombination of polymer radicals. With this aim the spin exchange was investigated [4.40] between stable macroradicals formed by binding nitroxide radicals to the end of the polymeric chain of polydimethyl siloxane and polyethylene oxide (Table 4.11). Simultaneously, with the help of the capillary method the diffusion coefficients of macroradicals D were measured. The experimental values of K_e were found to be in good agreement with the Smolukhovsky equation

$$K_e = 16\pi r_e D \,, \tag{4.6}$$

assuming r_e to be about the size of the paramagnetic center. This result means that the collisions of radical centers are limited by the diffusion mobility of the whole macromolecule.

In order to answer the question to what extent the results obtained by means of spin exchange can be applied to chemical reactions the data available in literature on the recombination of polymer radicals were analyzed [4.40] and the rate constants of recombination and those of spin exchange were compared. The two kinds of constants (normalized to the same viscosity) are found to be described by the same dependence on the mean degree of polymerization $\bar{P}$ (Fig.4.11)

$$K \sim (\bar{P})^{-0.9 \pm 0.1}$$

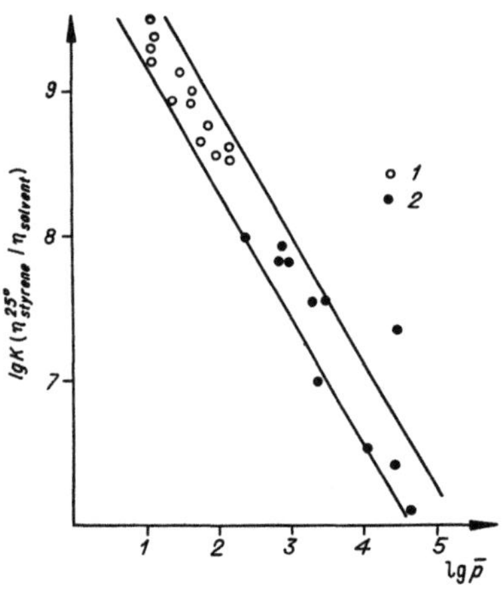

Fig. 4.11. The dependence of the rate constants of spin exchange (1) and of recombination (2) for macroradicals on the degree of polymerization. The constants are normalized to the viscosity of styrene at 25°C [4.40]. The rate constants for both processes are seen to follow the same dependence

Table 4.11. Spin exchange between nitroxide radicals attached to polymeric chain (styrene, 25 °C) [4.40]

Polymeric chain	$\bar{P}$ of polymer	K_e [l/mol · s]	D [cm^2/s]	r_{ef} [a] [Å]
Nonattached radical	1	$4.1 \cdot 10^9$	$1.8 \cdot 10^{-5}$	1.2
Polydimethyl siloxane	15	$2.8 \cdot 10^9$	$3.3 \cdot 10^{-6}$	4.6
Polydimethyl siloxane	29	$1.4 \cdot 10^9$	$3.0 \cdot 10^{-6}$	2.4
Polydimethyl siloxane	70	$7.4 \cdot 10^8$	$2.6 \cdot 10^{-6}$	1.7
Polyethylene oxide	140	$3.6 \cdot 10^8$	$1.8 \cdot 10^{-6}$	2.0

[a] Calculated from (4.6).

This regularity is true for so-called good solvents involving no such complicating effects as aggregation, etc. However, spin exchange is expected to enable one to model complex situations as well, because the efficiencies of the recombination and of the spin exchange of radicals are shown to be approximately the same [4.40].

4.1.3 Study of Intramolecular Collisions of Paramagnetic Fragments

In the chemistry of large molecules the intramolecular collisions of their functional groups may be just as important as the intermolecular collisions. To study the dynamics of these intramolecular collisions spin exchange in biradicals can be used.

At present there is a large number of publications on the spin exchange in biradicals (see, e.g, [4.41] and the appropriate chapters in [4.19,42]). As follows from the evidence presented therein the exchange interaction between the paramagnetic centers of a biradical may take place both via a bridge of chemical bonds between them and through collisions of these centers as these move diffusionally relative to one another. The latter mechanism prevails in cases when paramagnetic fragments are linked by a comparatively long flexible chain of bonds. It is only this kind of biradical that will be considered here. All the quantitative data available in the literature on the rates of the intramolecular collisions in biradicals were obtained from the analysis of the exchange narrowing of biradical spectrum lines (see the theory in Sect.2.5). Let us discuss a few typical examples of this kind of research.

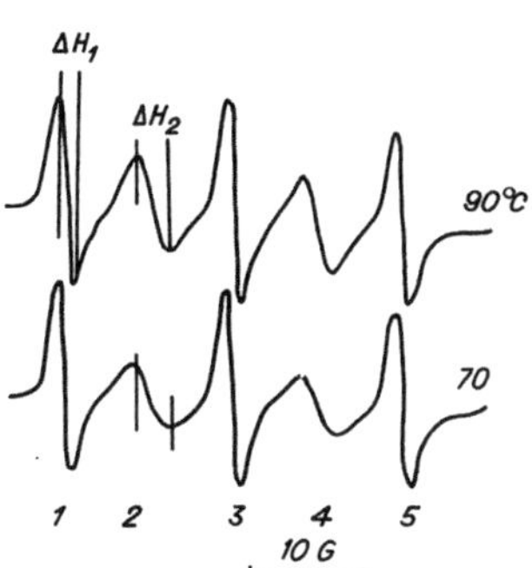

Fig. 4.12. The ESR spectrum of biradical I in dimethyl sulfoxide [4.43]. ΔH_K is the width of the k-th component between the points of the maximum slope. Exchange narrowing of components 2 and 4 takes place as temperature increases

A number of long-chain biradicals have been studied [4.43] whose structures are indicated in Table 4.12. As an example Fig.4.12 represents the ESR spectra of biradical 1 for two temperatures. As temperature increases one ob-

Table 4.12. Values of $\tau_0^{(0)}$ and E for long-chain biradicals

Biradical		Solvent	$-\lg \tau_0^{(0)} \pm 1$	E ± 0.6 [kcal/mol]
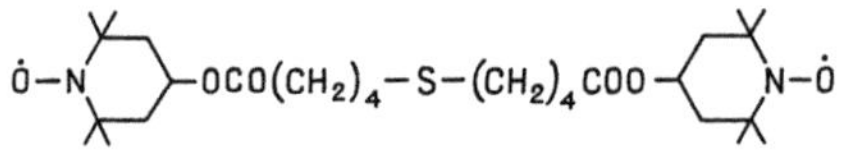	I	Benzene	11.5	3.2
		Toluene	11.7	3.4
		Dimethylsulfoxide	12.2	4.2
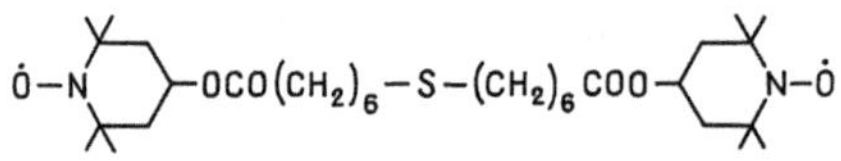	II	Dimethylsulfoxide	11.7	3.7
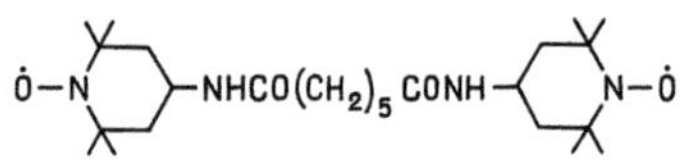	III	Benzene	12.1	3.5
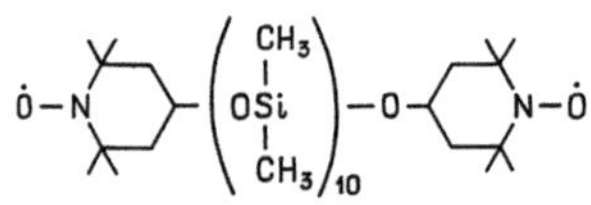	IV	Dimethylsulfoxide	11.7	2.9

serves exchange narrowing of components 2 and 4, i.e., a fast modulation of the exchange interaction occurs by intramolecular motion. In this case according to Sect.2.5.6 the exchange contribution to the widths of components 2 and 4 is described by the formula

$$1/T_2 = a^2\tau_0/4 \tag{4.7}$$

assuming that here, just as for collisions of nitroxide radicals in solution, the condition of strong exchange is fulfilled. Here τ_0 is the average time between repeated contacts of paramagnetic centers of a biradical giving rise to exchange interaction between them.

For the Lorentzian line shape from (4.7) one can obtain the following expression which is very convenient for practical calculations:

$$a\tau_0 = 3.46(\Delta H_2 - \Delta H_1)/\Delta H_{hf} \quad . \tag{4.8}$$

Here a and ΔH_{hf} represent the constant of hyperfine interaction measured in rad/s and Gauss, respectively; ΔH_2 and ΔH_1 are the widths of lines 2 and 1 measured between the points of the maximum slope.

The temperature dependences of $a\tau_0$ for the studied biradicals are given in Fig. 4.13 while Table 4.12 summarizes the parameters $\tau_0^{(0)}$ and E of Arrhenius plots

$$\tau_0 = \tau_0^{(0)} \circ \exp(E/RT) \quad . \tag{4.9}$$

From Table 4.12 it is seen that the preexponential factor $\tau_0^{(0)}$ for every system has the order of 10^{-11} - 10^{-12} s while the activation energies amount to 3-4 Kcal/mol. A typical value of time τ_0 between two collisions at room temperature is $\sim 10^{-9}$ s. These values of time and of activation energy E are in good qualitative agreement with the diffusion model of intramolecular collisions of paramagnetic fragments. Indeed, the values of E are close to those typical of the diffusion processes and rotations around ordinary bonds in liquids. The time τ_0 also coincides within an order of magnitude with the time of diffusion to distances of about 10-20 Å characterizing the size of a region in which a relative intramolecular diffusion of paramagnetic fragments of a biradical may occur.

In conclusion let us consider some peculiarities of the ESR spectra of long-chain biradicals that go beyond the range of effects described by the simplest double-conformation model of collisions discussed in Sect.2.5. For example, for biradical IV (see Table 4.12) the ESR spectrum corresponds to the superposition of two ESR spectra from the two states of a biradical [4.43]. The spectrum corresponding to one state was typical of fast exchange, the one fitting the second state was the usual triplet spectrum that characterized slow exchange. The ratio of the integral intensities of the two spectra was close to 1:1. To account for this fact, a triple-conformation model was suggested for a biradical in solution. Two conformations (A) with $J_A \gg a$ and (B) with $J_B \ll a$ are in the state of fast exchange with each other while the third (C) with $J_C \ll a$ exchanges with the former two rather slowly. The time of transition between conformations A and B is about $4 \circ 10^{-10}$ s, which again is within an order of magnitude equal to the diffusion times in such systems. The lifetime of conformation C is at least 30 times longer,

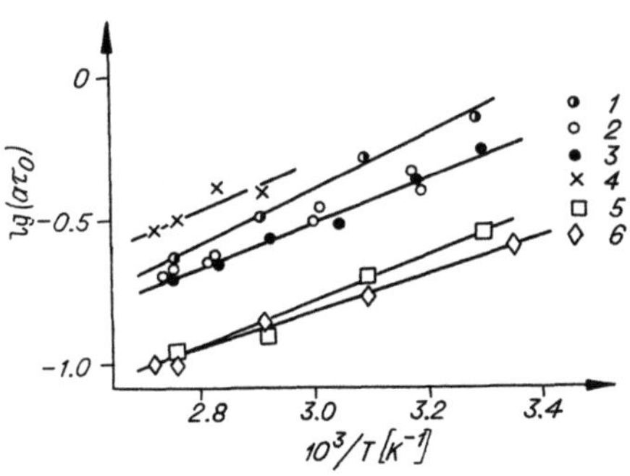

Fig. 4.13. The dependence of the parameter $a\tau_0$ on temperature in Arrhenius coordinates for biradicals I-IV [4.43]. 1: I in dimethyl sulfoxide; 2: I in benzene; 3: I in toluene; 4: II in dimethyl sulfoxide; 5: III in benzene; 6: IV in dimethyl sulfoxide

since otherwise the slow exchange condition would not be fulfilled. Thus, the transition between conformations A, B and conformation C takes place within times considerably exceeding those of the free diffusion of paramagnetic fragments. This appears to be the consequence of restrictions imposed on diffusion by a chain of chemical bonds separating the fragments.

To find the time τ_0, in many publications the difference $\Delta H_2 - \Delta H_1$ was determined not by direct measurement, but indirectly by measuring ΔH_1 and the ratio I_1/I_2 of the peak intensities of components 1 and 2. ΔH_2 was then calculated with the help of the equation

$$I_2 \cdot (\Delta H_2)^2 = 2I_1 \cdot (\Delta H_1)^2 \quad . \qquad (4.10)$$

This equation is valid for the first derivative of ESR lines with similar shapes. The above data make it necessary to be careful when dealing with the results of these measurements. Indeed, under the conditions of the superposition of a five-component spectrum from fast exchanging conformations, and of a three-component spectrum from slowly exchanging ones (4.10) loses its validity. Therefore the values of ΔH_2 and τ_0 obtained with its help may turn out incorrect. The situation becomes especially complicated if there is a superposition of more than two spectra. A biradical was studied, for example, whose ESR spectrum represents a superposition of signals from five different conformations [4.44].

Thus, spin exchange between specially introduced paramagnetic particles - spin probes - makes it possible to obtain unique information on the kinetics and mechanism of inter- and intramolecular collisions in complex heterogeneous systems including collisions in solutions of polymers and in polymeric matrices. The spin exchange method may be used to quantitatively study the dependence of the rate of molecular collisions on such factors of importance for chemical kinetics as composition of solvents, spatial distribution of reagents, etc.

4.2 Estimation of the Energy of Exchange Interaction During Collisions

In the preceding subsection examples were given of systems in which the situation of strong spin exchange is realized. From these data one can obtain the lower estimate for the value of the exchange integral of colliding paramagnetic particles. Thus for collision of free radicals the case of strong exchange requires the fulfilment of the condition $J\tau_c > 1$, and, hence, the exchange integral must be larger than $1/\tau_c$. Estimates of τ_c according to (2.120) or (2.121) show the typical value of τ_c to be of the order of 10^{-11} s. Therefore in the case of strong exchange between radicals we obtain $J > 10^{11}$ rad/s. Similar estimates in the strong exchange situation can also be made for collisions of a radical with a complex ion.

In the region of weak spin exchange the rate constant depends directly on the energy of exchange interaction, i.e., it contains direct information on magnitude of the exchange integral and, hence, on the overlapping of the orbitals of particles colliding in a solution. As is known, the overlapping of electron orbitals may also play a decisive role for some other elementary processes, e.g., for electron transfer reactions and for the quenching of triplet molecules. With respect to these processes spin exchange may be regarded as a modelling process making it possible to obtain data on the overlapping of partners orbitals and to establish the dependence of this overlapping on the chemical structure of the particles.

The study of weak spin exchange is also of interest because of the following. It is in this region that theory predicts specific dependences of the constant K_e on various parameters (τ_c, T_1, J) and, hence, can be subjected to critical verification. In particular, one can verify the consequences that follow from the Owen equation (2.11) establishing a relation between the exchange integral and the distribution of spin density over paramagnetic particles

Unfortunately, there are relatively few cases described in the literature where the occurrence of weak exchange is proved with confidence. The numerous examples discussed in Sect.4.1 show the strong spin exchange to be often observed both for collisions between radicals and for those between a radical and a paramagnetic complex. As follows from the theoretical equations reported in Chap. 2 there are, in principle, the following ways for transition to the situation of weak spin exchange.

1) Shortening of collision time by reducing the parameter η/T. For this purpose one can select solvents with low viscosity or raise the temperature of the solution.

2) Shortening of collision time by taking similarly charged particles.

3) Reduction of the value of the exchange integral by decreasing the spin density on the external atoms, accessible for collisions, of a paramagnetic particle.

4) Shortening of the relaxation time of one of the paramagnetic particles so that the condition T_1, $T_2 < \tau_c$ is fulfilled.

Experimental data on weak exchange are summarized below and crude estimates of the values of exchange integrals are given.

4.2.1 Free Radicals

The situation of weak spin exchange in the case of free radicals is rather unlikely since in these compounds atoms with appreciable spin density are easily accessible in collisions, so the exchange integrals must be comparatively large. To be convinced of this, take into consideration that according to theoretical estimates (see Sect.2.1) the value of the exchange integral during a Van der Waals contact of two atoms, each carrying a unit spin density, amounts to $\sim 10^{14}$ rad/s. Since the typical time of collision in a liquid is of the order of 10^{-11} s, we obtain $J\tau_c \approx 10^3 \gg 1$. Thus the condition of strong exchange is well fulfilled in a typical situation.

In the literature only one example has been described that was interpreted as a case of weak exchange between uncharged free radicals. That was the spin exchange in low-viscosity solutions of the DTBN radical. In Sect.4.1 data were presented showing that at viscosities $\eta > 0.3$ cP the spin exchange between these radicals is strong. To obtain a weak exchange situation measurements were made [4.7,8] in solvents with a very low viscosity such as liquid propane and n-pentane at elevated temperatures. To change viscosity the external pressure was also varied [4.8]. Qualitatively, the results of two groups of authors [4.7,8] agree well. The exchange broadening of the ESR lines in both solvents was found to be a linear function of T/η at high viscosities as is the case for strong exchange. At sufficiently large values of T/η, however, a deviation from linearity occurs, which is indicative of a transfer from the strong to the weak exchange case. As an example Fig.4.14 presents the data on the exchange broadening of the DTBN lines in propane [4.8]. Much attention was given in this case to the correct exclusion of various possible errors in measuring the exchange broadening.

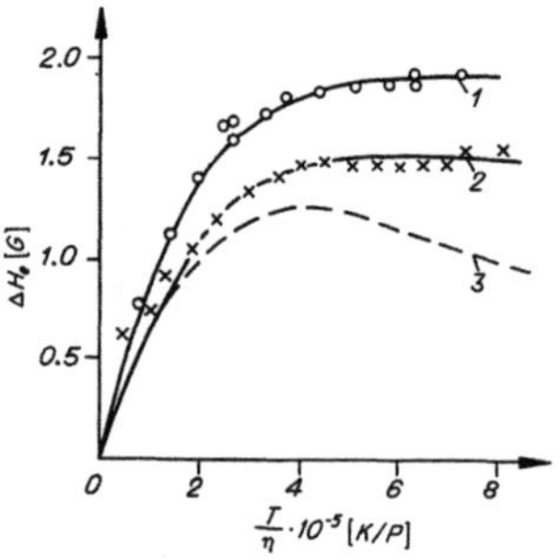

Fig. 4.14. The dependence of the exchange broadening ΔH_e on T/η for a solution of DTBN in propane [4.8]. Concentration of DTBN [mol/l]: 1 - $1.8 \cdot 10^{-3}$; 2 - $1.2 \cdot 10^{-3}$. Curve 3 is constructed in accordance with (4.11). The experimental curves are seen to reach plateau rather than to exhibit maximum as predicted by the theory

The peculiarity of the curves depicted in Fig.4.14 is that in the region of large T/η interpreted as a region of weak exchange the broadening remains constant rather than dropping as predicted by the Currin formula (2.144) for the efficiency of spin exchange between radicals

$$p = (1/2)J^2\tau_c^2/(1 + J^2\tau_c^2) \quad . \tag{4.11}$$

To account for this peculiarity a suggestion was made [4.7] that an elementary act of spin exchange is similar to a monomolecular chemical reaction in a cage of the solvent. Then, just as in the kinetic region of an ordinary chemical reaction, the rate constant of spin exchange and, hence, the exchange broadening must be independent of T/η. Such a suggestion is, however, in contradiction with the basic principles of spin exchange theory which is in good agreement with all the available experimental data (see also the discussion of this problem in Sect.2.5.4). Therefore, an attempt was made [4.8] to explain the dependence of K_e on T/η within the framework of (4.11). It was suggested [4.8] that the collision time of molecules in a liquid may be estimated by the formula

$$\tau_c = \pi \cdot b\lambda^2(\eta/kT) \quad , \tag{4.12}$$

where b is the effective hydrodynamic radius of a paramagnetic particle and λ is the length of a diffusion jump. The jump length λ was assumed to increase with temperature due to the growth of the free volume. To describe this dependence use was made of the formula

$$\lambda_T^3 = \lambda_0^3(\rho_0/\rho_T) \quad , \tag{4.13}$$

where ρ_T and ρ_0 are the liquid densities at a temperature T and at the freezing point, respectively; λ_T and λ_0 are the corresponding jump lengths. The above assumption can be readily seen to lead to a weakening of the dependence of τ_c on η/T. As a result, the entire experimental curve has been described fairly well [4.8] by choosing appropriate parameters. The exchange integral was found to be $J = 6 \cdot 10^{10}$ rad/s. This value appears to be somewhat too low because too high a value for the collision time τ_c is chosen. Within the jump model of passing the region of interaction the time τ_c must be two times shorter than what follows from (4.12) [cf with (2.120)].

One can also propose another simple qualitative explanation of the observed dependence of K_e on T/η. As has already been noted, the distribution of spin density over the external atoms of DTBN is extremely inhomogeneous.

Therefore the contacts of various atoms may result in different values of the exchange integral, i.e., the integral is anisotropic. As the viscosity gets lower the conditions of strong exchange, though violated for small exchange integrals, will persist for large ones. In this situation the region intermediate between strong and weak exchange will clearly be far wider than predicted by (4.11) with a single value of the exchange integral (see Sect. 2.5.4). We shall show below that the different values of exchange integrals may, e.g., correspond to the contacts of the atoms of NO ... ON and of NO ... HC. For such a model the exchange integral obtained above fits the contacts NO ... HC or long-distant contacts NO ... ON.

4.2.2 Spin Exchange with Metal Complexes

The conditions of weak exchange can be realized far more readily if at least one of the collision partners is a metal complex. Firstly, in this case one may essentially decrease the spin density on the external atoms of the complex by selecting proper ligands. Secondly, the efficiency of exchange may be lowered by selecting complexes with short relaxation times T_1, T_2.

To realize the first possibility, spin exchange with the Co(II) complexes formed by $Co(ClO_4)_2$ in alcohols and other organic solvents have been investigated [4.23,24]. The rate constants of the spin exchange of these complexes were measured with the radicals 2,2,6,6-tetramethyl-4-oxopiperidine-1-oxyl (TANONE) and α,α-diphenyl-β-picryl-hydrazyl (DPPH) as well as with the cation of bisdiphenylchromium (BDC). Special care was taken to remove any traces of water since the entrance of water molecules into the coordination sphere would essentially increase the efficiency of spin exchange. It should be remembered in this connection that according to the data of the preceding section the spin exchange of $Co(H_2O)_6^{2+}$ with TANONE is strong.

As was expected, the screening of the paramagnetic cobalt atom by organic molecules on whose external atoms the spin density is low leads to an essential decrease in the rate constants of spin exchange in comparison with the diffusion ones (Table 4.13). Temperature measurements on some systems have shown this decrease to be due precisely to the transition to weak exchange. Thus the rate constants of spin exchange with TANONE in acetone and dimethylsulfoxide are independent of temperature, while in methanol and other alcohols the temperature dependence is far less pronounced than should be expected for the diffusion constants (Fig.4.15). It is surprising that there is no increase in the screening effect of alcohols as the alkyl group gets longer. Detailed investigations of the electric conductivity and of NMR spectra have

Table 4.13. Rate constants of spin exchange between free radicals and complexes formed on dissolving of $Co(ClO_4)_2$ in organic liquids (20 °C) [4.23,24]

Solvent	$K_e \cdot 10^{-9}$ [l/mol · s]		
	TANONE	BDC	DPPH
CH_3OH	0.42	-	0.06
C_2H_5OH	0.33	0.01	0.01
C_3H_7OH	0.3	0.03	0.01
C_4H_8OH	0.3	0.05	0.01
sec-$C_8H_{17}OH$	0.5	-	0.16
CH_3COCH_3	0.15	0.04	0.02
CH_3SOCH_3	0.27 [a]	-	-

[a] Measured at 25 °C .

shown [4.23,24] that this might be connected with the coordination of ClO_4^- anion by the cobalt ion in heavier alcohols. Therefore the collisions of TANONE with ClO_4^- atoms, rather than with alcohol atoms, probably make the main contribution to K_e.

Table 4.14 summarizes the experimental data for processes in which the transition into the region of weak exchange was provided by shortening the relaxation time of the metal atom in the complex. Indeed, the above complexes may be divided into two groups. For the complexes VO(II), Cr(III), Mn(II), Fe(III), and Cu(II) the condition $T_{1,2} > \tau_c$ seems to be valid. Indeed, it is known that in the solutions of complexes of these metals the ESR spectra can be observed. This means that the width of the ESR spectrum which is determined by paramagnetic relaxation does not exceed the operation frequency ν_0 of the spectrometer: $1/(\pi T_2) < \nu_0$. Hence, by substituting $\nu_0 = 10^{10}$ s^{-1} (for the X-band ESR spectrometer) we obtain for the lower estimate of T_2 the value

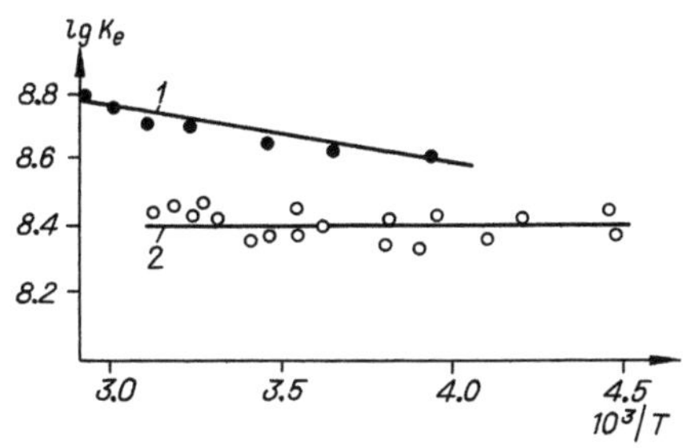

Fig. 4.15. The temperature dependence of the rate constant of spin exchange between TANONE and $Co(ClO_4)_2$ in methanol (1) and acetone (2) [4.23,24]

$T_2 > 3 \cdot 10^{-11}$ s which is of the same order of magnitude as the collision time in a liquid. In contrast, for the Co(II), Fe(II), Ni(II), and Mn(III) complexes one fails to observe any ESR spectra in solutions, so it seems that $T_{1,2} < \tau_c$ for them.

The analysis of the results obtained (see Table 4.14) shows that the rate constants of spin exchange for the complexes of the first group ($T_{1,2} > \tau_c$) are larger than those of the second ($T_{1,2} < \tau_c$). Therefore for at least the complexes with short relaxation times the exchange may be considered to be weak. As to the complexes with long relaxation times it cannot be established unequivocally on the basis of the data of Table 4.14 alone whether the spin exchange is strong or weak.

To confirm the conclusion that weak exchange takes place for acetylacetonates with $T_{1,2} < \tau_c$, a detailed study was made of the spin exchange of the complexes $Ni(acac)_2L_2$ with TANONE [4.23,24]. In the first series of experiments [4.24] the exchange integral was varied by changing the nature of the ligand L. In the case of weak exchange this must result in a change in the rate constant of spin exchange in accordance with the value of the spin density delocalized onto the atoms of the ligand. The greatest constant is observed for aniline where the spin density is delocalized through the π bonds of the aromatic ring nearly without attenuation [4.26,46]. Unlike aniline,

Table 4.14. Rate constants of spin exchange between free radicals and transition metals acetylacetonates in chloroform [4.45]

T_1 vs τ_c	Complex	$K_e \cdot 10^{-9}$ [l/mol · s]		
		TANONE	DPPH	BDC
$T_1 > \tau_c$	$VO(acac)_2$	1.6	-	-
	$Cr(acac)_3$	1.9	2.5	1.0
	$Mn(acac)_2$ [a]	3.0	-	1.6
	$Fe(acac)_3$	3.2	3.0	1.1
	$Cu(acac)_2$	2.2	2.3	1.2
$T_1 < \tau_c$	$Mn(acac)_3$	0.5	0.9	0.5
	$Co(acac)_2$	0.4	-	0.2
	$Ni(acac)_2$	0.22	0.21	0.18

[a] K_e was measured in pyridine.

Table 4.15. Rate constants of spin exchange between TANONE and $Ni(acac)_2L_2$ complexes in chloroform at 20 °C [4.24]

No.	Ligand	$K_e \cdot 10^{-9}$ [l/mol · s]
1	$C_6H_5NH_2$	1.00
2	$(CH_3)_2SO$	0.54
3	$C_5H_{10}NH$	0.39
4	$P(C_6H_5)_3$	0.22
5	$OP(C_6H_5)_3$	0.20
6	C_2H_5OH	0.22
7	C_5H_5N	0.22

pyridine is a σ electron ligand in a complex with $Ni(acac)_2$ and the spin density quickly attenuates as the distance from the metal increases [4.26,46]. Accordingly, the rate constant of spin exchange for $Ni(acac)_2Py_2$ is smaller. It should be noted that the independence of K_e on the nature of the ligand in systems 4-7 may be because the spin exchange is carried out in these cases through the atoms of acetyl-acetonate-ions rather than the ligands L.

In another experiment [4.23] the temperature dependences of the spin exchange rate constant were measured (Fig.4.16). For the complex $Ni(acac)_2$ with pyridine K_e is practically independent of temperature which corresponds to the case of weak exchange (for systems with $T_{1,2} < \tau_c$). When the ligand is aniline the dependence observed corresponds to strong exchange.

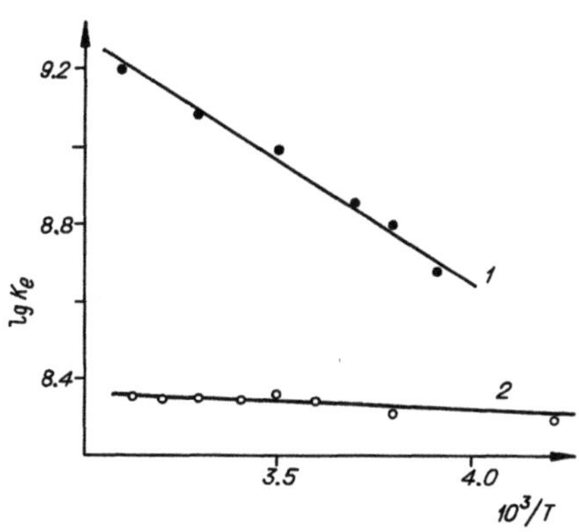

Fig. 4.16. The temperature dependence of the rate constant of spin exchange between TANONE and $Ni(acac)_2L_2$ in chloroform [4.23]. 1 - L aniline; 2 - L pyridine

The results discussed make it possible to estimate the value of the exchange integral for some of the systems studied. From (2.210) for the rate constant of spin exchange in the case of short times T_1, we have

$$J = \{[3/2S(S + 1)T_1\tau_c]K_e/fK_D\}^{1/2} \quad . \tag{4.14}$$

Substituting into this formula the expression for K_D and τ_c according to (4.4) and (2.121), respectively, we obtain

$$J = \{(3000/32\pi)(K_e/f)/[N_0R^2aT_1S(S + 1)]\}^{1/2} \quad , \tag{4.15}$$

where N_0 is the Avogadro number. As a rough approximation let us assume the steric factor f for all systems considered to equal one. As could be seen in Sect.4.1 for strong exchange this assumption is really fulfilled in many cases. For the particle radius and the thickness of the spherical layer in which spin exchange is effective we assume $R = 4 \cdot 10^{-8}$ cm, $a = 0.5 \cdot 10^{-8}$ cm. As a result we obtain the following approximate expression:

$$J = 2.5(K_e/T_1S(S + 1))^{1/2} \quad , \quad \mathrm{rad/s} \quad , \tag{4.16}$$

where T_1 is in s and K_e in 1/mol · s.

Table 4.16 lists estimates of the exchange integrals J made with (4.16) for systems where the electron relaxation time T_1 is known and the spin exchange can be assumed to be weak on the basis of the above discussion. The table also includes a few strong exchange systems with the estimated lower limit of the exchange integral. The condition of strong exchange

$$(2/3)S(S + 1)J^2T_1\tau_c > 1 \tag{4.17}$$

was used for this purpose where τ_c was calculated according to (2.121) with the parameters assumed above.

As follows from Table 4.16 and the results of the previous section, the exchange integrals for a radical-complex collision may be smaller than for a radical-radical collision. As to the magnitudes of J obtained for either kind of systems, they do not go beyond the limits of the interval covering two orders of magnitude ($5 \cdot 10^{11} - 5 \cdot 10^{9}$ rad/s). Undoubtedly this result is only a consequence of limitations of the experimental procedure. Indeed the methods described do not enable one to measure the exchange integrals larger than τ_c^{-1} or $(T_1 \cdot \tau_c)^{-1/2}$ and smaller than ones at which $K_e \lesssim 10^7$ 1/mol · s.

Table 4.16. Exchange integrals estimated from the data on spin exchange between radicals and complex ions with $T_1 < \tau_c$

Radical	Complex	$K_e \cdot 10^{-9}$ [l/mol · s]	T_1[s]		J [rad/s]
Weak exchange					
TANONE	$Ni(acac)_2Py_2$	0.22	$\sim 10^{-11}$ [a]		$8 \cdot 10^9$
TANONE	$Mn(acac)_3$	0.5	10^{-11}-10^{-12}	[4.47]	$7 \cdot 10^9 - 2 \cdot 10^{10}$
TANONE	$Co(CH_3OH)_6^{2+}$	0.42	$7 \cdot 10^{-13}$	[4.48]	$3 \cdot 10^{10}$
DPPH	$Co(CH_3OH)_6^{2+}$	0.06	$7 \cdot 10^{-13}$	[4.48]	$1.2 \cdot 10^{10}$
DPPH	$Co(C_2H_5OH)_6^{2+}$	< 0.01	$\sim 10^{-12}$ [b]		$< 4 \cdot 10^9$
Strong exchange					
TANONE	$Co(H_2O)_6^{2+}$	0.9	$\sim 10^{-12}$	[4.48]	$> 10^{11}$
TANONE	$Ni(acac)_2 \times (C_6H_5NH_2)_2$	1.0	$\sim 10^{-11}$ [a]		$> 6 \cdot 10^{10}$
TANONE	$CoPy_4Cl_2$	1.4	$\sim 10^{-12}$ [b]		$> 10^{11}$
BDC	$CoPy_4Cl_2$	2.4	$\sim 10^{-12}$ [b]		$> 10^{11}$

[a] T_1 is taken to equal $\sim 10^{-11}$ s, just as for $Ni(acac)_2$ complexes with alkyl amines [4.49].

[b] T_1 is taken to equal $\sim 10^{-12}$ s, just as in other octahedral·complexes.

4.2.3 Discussion of Results

The above results of research on weak exchange are in qualitative agreement with the Owen formula (see Sect.2.1)

$$J = J_{ab} \cdot \rho_a \rho_b / n_A \cdot n_B \ , \qquad (4.18)$$

where ρ_a and ρ_b are the spin densities on atoms a and b that are in close contact when two particles A and B collide in a solution; J_{ab} is the exchange integral between the atoms a and b at a Van der Walls distance between them; n_A and n_B are the numbers of unpaired electrons in A and B. In accordance with (4:18) the rate constants of the radical-metal complex spin exchange decrease with decreasing spin density in the periphery of the complex particle. Thus, the spin densities on the outermost atoms of ligands decrease in the order $Cl^- > H_2O > CH_3OH > C_2H_5OH$. The rate constants for weak spin exchange and the estimated exchange integrals decrease in the same sequence.

It is of interest to find out whether one can use (4.18) for the estimation of the exchange integrals during collisions in solutions. To answer this question one should compare the values of J estimated from spin exchange data with those calculated according to (4.18). Such a comparison, however, contains considerable elements of arbitrariness. Firstly, the above estimates of J from the experimental data are insufficiently accurate due to the uncertainty of various parameters. Secondly, the values of J_{ab} for different pairs of atoms needed for calculations according to (4.18) are known to be insufficiently reliable. The quantochemical calculations are only just being started (see Sect.2.1), and so far it is difficult to judge their accuracy.

Finally, the application of (4.18) to collisions in solutions also meets with some difficulties because the structure of a bridged group in solutions, contrary to that in solid paramagnetics, is not, as a rule, known beforehand. Indeed the colliding particles are oriented arbitrarily, the orientation itself changing randomly during the collision. Strictly speaking, it is not clear a priori which atoms of the colliding paramagnetic particles make the main contribution to the exchange interaction. Yet the decisive contribution to the exchange interaction is expected to be made by the contacts of atoms carrying sufficiently large spin densities. This circumstance allows one in many cases to make a reasonable suggestion as to the configurations of bridged structures that are the most effective for spin exchange.

Bearing in mind to above uncertainties it is reasonable to restrict oneself to discussing the applicability of (4.18) with an accuracy of about one order of magnitude. With this in view assume $J_{ab} \approx 2 \cdot 10^{14}$ rad/s irrespective of the nature of the colliding atoms (see the estimates of J_{ab} in Sect.2.1).

Let us begin by considering a collision of two DTBN radicals. The hydrogen and oxygen atoms of this radical that are attainable for collision carry spin densities $\rho_O \approx 0.6$ and $\rho_H \approx 2 \cdot 10^{-4}$ [4.19]. One can imagine three types of possible collision contacts of such radicals (Fig.4.17). The value of the exchange integrals for these situations, according to (4.18), are: $J_{O...O} = 7 \cdot 10^{13}$, $J_{O...H} = 2.4 \cdot 10^{10}$ and $J_{H...H} = 8 \cdot 10^{6}$ rad/s. Comparison of these values with the exchange integral evaluated [4.8] from spin exchange data ($6 \cdot 10^{10}$ rad/s) shows that the best agreement corresponds to contacts of the oxygen atoms of one radical with the hydrogen atom of the partner. The oxygen-oxygen contacts can also make a substantial contribution to the exchange broadening of ESR lines. Thus in the example considered the application of the Owen formula makes it possible to suggest the most probable structure of the intermediate complex providing exchange interaction in a collision.

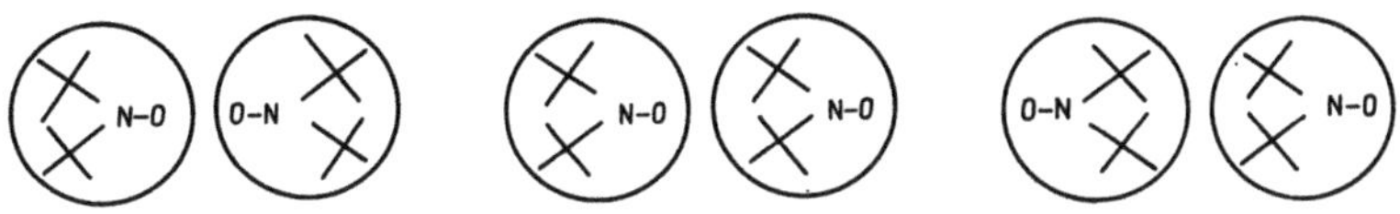

Fig. 4.17. Possible configurations of the exchange complex during a collision of two DTBN radicals

The results of a similar consideration for collision of radicals with complex compounds are presented in Table 4.17. Only those structures of exchange bridges which yield the greatest values of the exchange integral and are not forbidden sterically are included here. The values of spin densities

Table 4.17. Comparison of exchange integrals estimated from spin exchange data and by means of (4.18)

Radical	Complex	Assumed structure of exchange bridge	ρ_A	ρ_B	J [rad/s] according to (4.18)	J [rad/s] from Table 4.16
TANONE	$CoPy_4Cl_2$	NO ... ClCo	0.6	0.05	$2 \cdot 10^{12}$	$>10^{11}$
TANONE	$Co(H_2O)_6^{2+}$	NO ... HOCo	0.6	$\sim 10^{-3}$	$4 \cdot 10^{10}$	$>10^{11}$
		NO ... OCo	0.6	$\sim 10^{-2}$	$4 \cdot 10^{11}$	$>10^{11}$
TANONE	$Co(CH_3OH)_6^{2+}$	NO ... HCOCo	0.6	$8 \cdot 10^{-4}$	$3 \cdot 10^{10}$	$3 \cdot 10^{10}$
BDC	$CoPy_4Cl_2$	$=\underset{\mathrm{H}}{\overset{\mid}{\mathrm{C}}}$ ... ClCo	$2.5 \cdot 10^{-2}$	0.05	$8 \cdot 10^{10}$	$>10^{11}$
DPPH	$Co(CH_3OH)_6^{2+}$	$=\underset{\mathrm{H}}{\overset{\mid}{\mathrm{C}}}$ ... HCOCo	~ 0.1	$8 \cdot 10^{-4}$	$5 \cdot 10^{9}$	$1.2 \cdot 10^{10}$

on the atoms of paramagnetic particles which are necessary for estimating J are borrowed from original papers on hfi in the ESR and NMR of these compounds [4.26,46].

As seen from Table 4.17, the estimates according to the Owen formula are generally in agreement with the results of experiments on spin exchange. The chlorine atom provides, as a bridged group, large exchange integrals thus giving rise to strong exchange. In the case of spin exchange of TANONE with the cobalt aquo complex the strong exchange is probably achieved by contacts of the NO group with oxygen atoms of coordinated water. The coincidence of absolute values of J is also satisfactory in the cases where comparison is possible. The examples considered enable one to conclude that the Owen formula may be a useful tool in analyzing experimental data on spin exchange for it allows one to predict the trends in the variation of the exchange integral with the structure of interacting particles, and sometimes to establish a detailed mechanism of interaction of paramagnetic particles in a collision.

It is of interest to compare the results considered with the values of exchange integrals in systems where the structure of the group bridging the paramagnetic centers is known. Some paramagnetic metal oxides and salts, clusters containing two or more paramagnetic ions, crystals of organic radicals, and biradicals are examples of such systems. Evidence on exchange interactions in such systems can generally be obtained from the data on the temperature dependence of magnetic susceptibility or on the ESR spectra. The number of publications containing this kind of data is rather large. Here we shall cite only some typical examples (Table 4.18).

As seen from Table 4.18, in case the paramagnetic ions are separated by one or two intermediate atoms, the exchange integrals are large enough. If such a structure of the bridge is realized during collisions in solutions it must lead to strong exchange. This conclusion is in full agreement with the above results showing that the bridged chlorine atom ensures a strong exchange between a radical and a cobalt ion and that the lower limit of J is $J > 10^{11}$ rad/s.

Exchange integrals between free radicals in crystals (see Table 4.18) are also in qualitative agreement with the results on spin exchange. Even in the case when NO groups are not in close contact, the exchange integrals are sufficiently large ($\sim 10^{12}$ rad/s) and, hence, may ensure strong exchange. Thus, these data agree with the above suggestion that the contacts of the oxygen atom of the NO group of one radical with atoms less abundant in spin density

Table 4.18. Exchange integrals in some systems with a known structure of bridge group

System	Structure of bridge group	Method	J [rad/s]	Reference
NiO	Ni–O–Ni	MS [a]	$1.5 \cdot 10^{13}$	[4.50]
$KNiF_3$	Ni–F–Ni	MS	$9.1 \cdot 10^{12}$	[4.25]
$K_2(Ir, Pt)Cl_6$	Ir–Cl–Cl–Ir	MS	$1.5 \cdot 10^{12}$	[4.51]
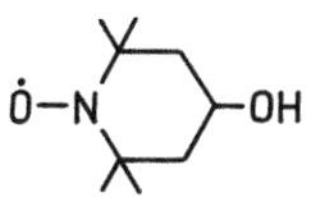	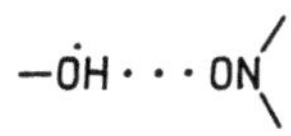	MS	$\sim 10^{12}$	[4.52]
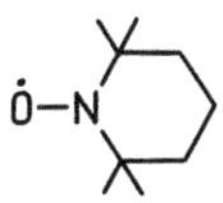	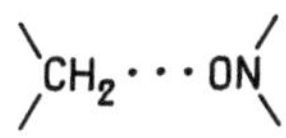	MS	$\sim 10^{12}$	[4.52]
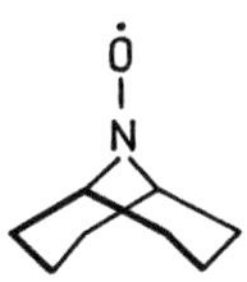	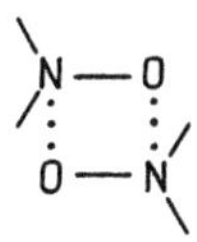	MS	$\gtrsim 10^{14}$	[4.52]
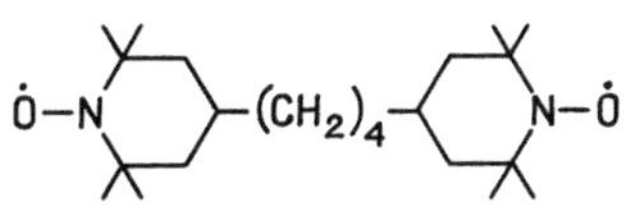	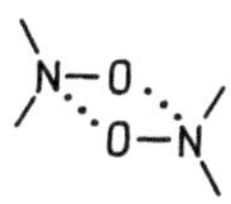	MS	$\sim 10^{13}$	[4.52]
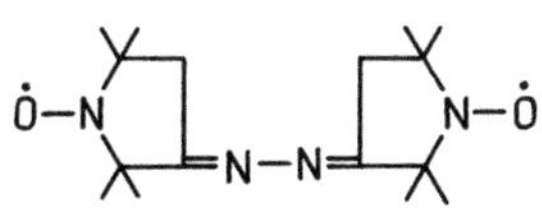	Intramolecular exchange	MS	$\sim 10^{12}$	[4.52]
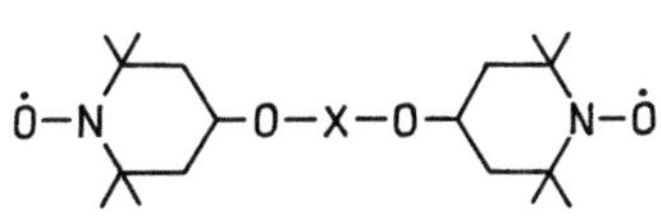 [b]	The same	ESR	$(1.8\text{-}2.5) \times 10^{9}$	[4.53]

[a] MS - magnetic susceptibility measurement; the exchange integrals were estimated from the maximum position of magnetic susceptibility as a function of temperature.

[b] X = 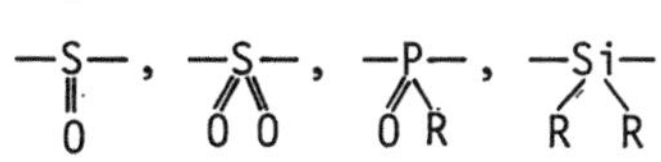 .

of another radical can contribute substantially to exchange interaction of nitroxide radicals in solution.

Data on the intramolecular spin exchange in biradicals (Table 4.18) are in qualitative agreement with the above evidence that the nature of the ligands separating the paramagnetic atom of metal from the free radical may essentially influence the efficiency of spin exchange.

Thus, the exchange integrals in bridged structures are generally in good agreement with the data obtained in investigating spin exchange. This shows additionally that the indirect exchange mechanism plays an important role in spin exchange in solutions.

4.3 Collisions of Charged Particles

The rate of chemical reactions between charged particles is known to be influenced by electrostatic repulsion or attraction between them. The quantitative evaluation of the contribution of this interaction to the rate constant for condensed media is a very complicated problem.
The theoretical calculation of this contribution presents problems because of the necessity of taking into account the spatial distribution of the charge over the reagents (in nonspheric particles), the possibility of associations of reagents with a counterion and the lack of reliable data on microscopic dielectric constant (ε_{micro}) of the medium in the vicinity of ions, etc. As to the experimental research on the rôle of electrostatic interactions in ionic reactions, it is complicated by the fact that their rate usually depends on a great number of factors each of which is difficult to estimate a priori so that it is extremely difficult to single out experimentally the electrostatic contribution to the rate constant.

The spin exchange between charged particles is influcenced by a much smaller number of factors. The study of spin exchange is therefore a convenient method of investigating the influence of electrostatic interactions on the efficiency of various processes in liquids. Especially convenient in this respect are the cases of strong exchange [$J\tau_c$, $J(T_1\tau_c)^{1/2} > 1$] when the rate of the process is controlled by diffusion.

Spin exchange was studied between charged particles themselves [4.10,17, 32,54-58] as well as between their associates containing counterions [4.9, 10,59-62]. Let us analyze these results in detail.

4.3.1 Experimental Results

Spin Exchange in Solutions of Anion Radical $(SO_3)_2NO^{2-}$

The broadening of the ESR lines of a doubly charged nitrosyl-disulfonate anion radical $(SO_3)_2NO^{2-}$ was measured [4.54] under the influence of paramagnetic ions and radicals with the formal charges +3, +2, +1, -1, -2, -3 (Fig.4.18). The Lorentzian shape of the initial ESR lines and their small width (~0.3 G in nondegassed water solutions) combined with a comparatively large hf splitting (~13 G) make it possible to measure the exchange broadening in a very wide range of concentrations (10^{-4} ÷ 10^{-1} mol/l). A certain drawback of nitrosyl-disulfonate anion radical is its instability in water solutions.

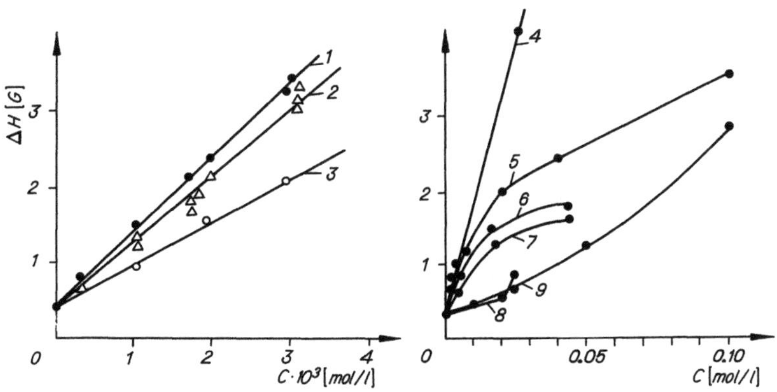

Fig. 4.18. The dependence of the line width of the $(SO_3)_2NO^{2-}$ ESR spectrum in aqueous solutions on the concentration of paramagnetic particles [4.54]. 1: $Cr(H_2O)_6^{3+}$, $Cr(en)_3^{3+}$, $Gd(H_2O)_n^{3+}$; 2: $Co(H_2O)_6^{2+}$, $Ni(H_2O)_6^{2+}$, $Cu(H_2O)_6^{2+}$; $Cr(NH_3)_5Cl^{2+}$; 3: $Cr(SCN)_2(en)_2^{+}$; 4: $Cr(en)\ (C_2O_4)_2^{-}$; 5: $Ce(H_2O)_n^{3+}$; 6: $Nd(H_2O)_n^{3+}$; 7: $Pr(H_2O)_n^{3+}$; 8: $(SO_3)_2NO^{2-}$; 9: $Cr(CN)_6^{3-}$. Deviations from linearity result from the screening of Coulomb interactions by counterions

In Fig.4.18 the dependences of the line width of $(SO_3)_2\ NO^{2-}$ on the concentration of added paramagnetics C [4.54] are seen to be linear only at small C and to lose linearity at large C. In the latter case the corresponding curves deviate upwards in case the charge of the broadening agents is of the same sign (-) as that of the indicator, and downwards if the sign is opposite (+).

Table 4.19. Initial slopes of broadening vs concentration curves and rate constants of spin exchange between $(SO_3)_2NO^{2-}$ and various ions

No.	Ion	H_e/C [G · l/mol]	$K_e \cdot 10^{-9}$ [l/mol · s]	$fK_D \cdot 10^{-9}$ [l/mol · s]
1	$Cr(H_2O)_6^{3+}$	1010	15.3	24.5
2	$Cr(en)_3^{3+}$	1010	15.3	24.5
3	$Gd(H_2O)_n^{3+}$	1010	15.3	23.2
4	$Co(H_2O)_6^{2+}$	920	14.0	14.0
5	$Ni(H_2O)_6^{2+}$	940	14.3	14.3
6	$Cr(NH_3)_5Cl^{2+}$	950	14.5	23.2
7	$Cu(H_2O)_6^{2+}$	940	14.3	28.6
8	$Cr(en)_2(SCN)_2^+$	580	8.8	14.1
9	$Cr(en)(C_2O_4)_2^-$	150	2.3	3.68
10	$(SO_3)_2NO^{2-}$	10	0.23	0.46
11	$Cr(CN)_6^{3-}$	11	0.17	-
12	$Ce(H_2O)_n^{3+}$	160	2.4	-
13	$Nd(H_2O)_n^{3+}$	100	1.5	-
14	$P(H_2O)_n^{3+}$	80	1.2	-

Table 4.19 summarizes the slopes $\Delta H_e/C = (\Delta H - \Delta H_0)/C$ of the initial part of the curves in Fig.4.18 and the spin exchange rate constants K_e calculated from them. Comparing these values of K_e with the contribution of dipole-dipole interaction to the concentration broadening at the viscosity $\eta = 1$ cP (see Sect.3.3) we see that the observed broadening of the lines of $(SO_3)_2NO^{2-}$ at least by ions 1-10 is mainly due to exchange rather than dipole-dipole interaction in contrast to what was proposed originally [4.54].

The data of Table 4.19 and Fig.4.18 reveal a distinct influence of electrostatic interactions on the rate constant of spin exchange. Indeed, for the exchange between $(SO_3)_2NO^{2-}$ and oppositely charged ions 1-8 (Table 4.19) the constants K_e are far larger than for that between $(SO_3)_2NO^{2-}$ and similarly charged ions 9-11.

The deviation of the cations of rare earth elements 12-14 from this regularity appears to be due to their very short time of electron spin relaxation.

As a result, the situation of weak exchange $[J(T_1 \cdot \tau_c)^{1/2} \ll 1$ and $p \ll 1]$ is realized. For the collisions of these ions with $(SO_3)_2NO^{2-}$ such an explanation is supported by a large value of K_e for another rare earths cation $Gd(H_2O)_n^{3+}$, which seems to have a structure similar to that of ions 12-14, but in contrast to these has a longer electron spin relaxation time $T_1 \approx 10^{-8}$ s [4.63].

Note that the shape of curves 5-9 of the dependence of ΔH on C in Fig.4.18 also agrees with the idea of the influence of electrostatic interactions on K_e. Indeed, at great concentrations the effective charges of ions become lower due to the association with counterions as well as to the screening effect of the ionic atmosphere created by the counterions. As a result, the influence of electrostatic interaction must get weaker and the rate constants of exchange betweeen similarly charged particles must increase while those between oppositely charged ones must decrease. In consequence, the $\Delta H = f(C)$ curves for the exchange between similarly charged particles at large concentrations must bend upwards and for that between oppositely charged ones, downwards, which is indeed observed in the experiment.

For the exchange of $(SO_3)_2NO^{2-} + (SO_3)_2NO^{2-}$ and $(SO_3)_2NO^{2-} + Co(H_2O)_6^{2+}$ the dependence of the initial slopes of the curves $\Delta H = f(C)$ on the concentration of diamagnetic electrolyte KCl was studied (Fig.4.19). As KCl is added the values of $\Delta H_e/C$ increase for the exchange between similarly, and decrease for that between oppositely, charged particles. Such a change in $\Delta H_e/C$ is also accounted for by a decrease in the effective charges of paramagnetic ions due to their association with counterions and/or the formation of a charge-screening ionic atmosphere of counterions. The plateau on the curves in Fig.4.19 at high concentrations of KCl appears to correspond to a complete suppression of the influence of electrostatic interactions on spin exchange.

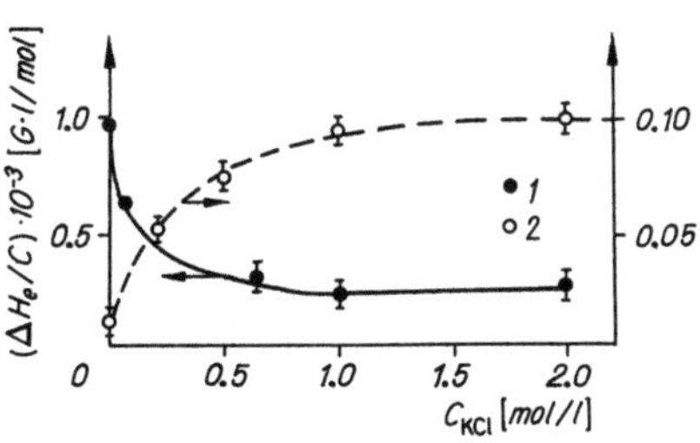

Fig. 4.19. The dependence of the slope of the initial parts of $(SO_3)_2NO^{2-}$ ESR lines broadening curves on the concentration of KCl [4.54]. 1: Exchange with $Co(H_2O)_6^{2+}$; 2: Exchange with $(SO_3)NO^{2-}$

The large values of K_e (10^{10} l/mol · s) for processes involving cations 1-8 (see Table 4.19) suggest that their collisions with $(SO_3)_2NO^{2-}$ create the situation of strong exchange. Note that for spin exchange between cation 4 and a TANONE radical which is close in structure to $(SO_3)_2NO^{2-}$ but uncharged the fulfillment of the strong exchange condition was proved by direct experiments [4.23].

Anions 9-10 are close in size to cations 1-8 and, just as these, must have a high enough spin density on atoms attainable for colliding with the oxygen atom of the NO group of the $(SO_3)_2NO^{2-}$. At the same time anions 9 and 10 have a large value of T_1, and the duration of their collision with $(SO_3)_2NO^{2-}$ cannot be shorter than the electron spin relaxation time for $Co(H_2O)_6^{2+}$ ($T_1 \approx 10^{-12}$ s [4.48]). Therefore, since the exchange for $(SO_3)_2NO^{2-}$ + $Co(H_2O)_6^{2+}$ is strong, then the exchange for $(SO_3)_2NO^{2-}$ + $(SO_3)_2NO^{2-}$ and $(SO_3)_2NO^{2-}$ + $Cr(en)_2(C_2O_4)^-$ must be strong too. It is difficult to say whether the exchange involving anion 11 is strong or weak since the time T_1 for this compound is unknown, dipole broadening can not be excluded also in this case.

In the case of strong exchange from the values of K_e one can calculate the product of the diffusion rate constnat K_D of collisions between paramagnetic particles by the steric factor using the formulae

$$fK_D = K_e/p_{max} = [3(S_B + 1/2)^2/2S_B(S_B + 1)]K_e \qquad (4.19)$$

for the case of broadening agents with a long T_1, and

$$fK_D = K_e \qquad (4.20)$$

for the case of broadening agents with a short T_1. These formulae correspond to the data listed in Table 4.1 for the spin of the indicator $S_A = 1/2$. The values thus obtained are summarized in Table 4.19.

The spin exchange between the $(SO_3)_2NO^{2-}$ anion radicals themselves has also been studied in [4.10,55]. The exchange frequency was determined not only from the broadening but also from the shift of the ESR lines [4.55]. Excellent agreement was found between the two techniques of measurement, the exchange nature of the broadening of the ESR lines of $(SO_3)_2NO^{2-}$ being thereby confirmed. As distinct from [4.54] all these experiments were carried out in aqueous solutions with 0.05M of K_2CO_3 added. These additives ensured a higher stability of the $(SO_3)_2NO^{2-}$ anion radicals which are stable in an alkaline medium. According to the data presented in Fig.4.19 under these conditions the screening of the charge of $(SO_3)_2NO^{2-}$ by counterions must already be considerable. It was found [4.55] that the dependence of exchange frequency on concentration is well described by

$$\omega_e = b \cdot C^m \quad . \tag{4.21}$$

The values of the coefficient b and of the exponent m obtained from different series of experiments are summarized in Table 4.20. Note the deviation of m from 1 which indicates that the rate constant of spin exchange increases with the concentration of anion radicals as

$$K_e = bC^{m-1} \quad . \tag{4.22}$$

Such an increase appears to be attributed to further screening of the charge of $(SO_3)_2NO^{2-}$ as its concentration increases. At low concentrations of paramagnetic particles $C^{m-1} \to 1$ and one may simply take $K_e = b$.

Table 4.20. Parameters b and m for spin exchange between $(SO_3)_2NO^{2-}$ anion radicals in 0.05 M K_2CO_3

Parameter measured	Sample treatment	$b \cdot 10^{-9}$ $(l/mol)^m \cdot s^{-1}$	m	References
Broadening	Degassed	1.84	1.07 ± 0.02	[4.55]
Shift	Degassed	1.29	0.98 ± 0.02	[4.55]
Broadening	Nondegassed	1.61	1.06 ± 0.02	[4.55]
Shift	Nondegassed	1.55	1.03 ± 0.02	[4.55]
Broadening	Deoxygenated by nitrogen bubbling	1.96	1.17 ± 0.02	[4.10]

The results discussed were reproduced in a subsequent investigation [4.10] where not only the values of b and m were determined in a 0.05 M solution of K_2CO_3 (see Table 4.20), but also the dependence was studied of K_e on the concentration of further added diamagnetic electrolytes K_2CO_3 and K_2SO_4 (up to $C \approx 0.7$ M). In accord with [4.54] K_e was found to increase, as the electrolytes were added, tending to a certain limit. Thus, at a 0.024 M concentration of $(SO_3)_2NO^{2-}$ and $t = 24^\circ C$ $K_e = 1.99 \cdot 10^9$ l/mol $\cdot$ s for 0.05 M of K_2CO_3, $2.98 \cdot 10^9$ - for 0.05 M of K_2CO_3 + 0.25 M of K_2SO_4, and $3.10 \cdot 10^9$ - for 0.05 of K_2CO_3 + 0.56 of K_2CO_4 [4.10].

The maximum value of $K_e = 3.19 \cdot 10^9$ l/mol · s obtained [4.10] at the highest concentration of diamagnetic electrolyte differs by 40% from the corresponding value of $K_e = 2.3 \cdot 10^9$ l/mol · s obtained in [4.54]. This is probably accounted for by an error in the concentrations of $(SO_3)_2NO^{2-}$ in [4.54]. EASTMAN et al. [4.10] and JONES [4.55] controlled the concentration of this unstable particle spectrophotometrically right in the solution while PEARSON and BUCH [4.54] did not seem to do so. Note that the error in the concentration of $(SO_3)_2NO^{2-}$ [4.54] does not affect, however, the results of measuring the rate constants of the exchange of $(SO_3)_2NO^{2-}$ ions with ions of another kind.

The strength of the spin exchange for the $(SO_3)_2NO^{2-} + (SO_3)_2NO^{2-}$ process was also studied [4.10]. It was found that as temperature rises the broadening ΔH_e of the ESR lines of $(SO_3)_2NO^{2-}$ increases linearly with T/η. The straight line, however, does not pass through the origin as must be the case for strong exchange. The slope decreased with decrease in the concentration of the added diamagnetic electrolyte. On the basis of these data it was suggested that the exchange in this system corresponds to the intermediate region ($J\tau_c \approx 1$). However, in view of the discussion of the data listed in Table 4.19, violation of the conditions of strong exchange for $(SO_3)_2NO^{2-}$ seems unlikely. Apparently, the effect observed [4.10] can be accounted for by the fact that at different concentrations of electrolyte the temperature dependences of the factor Q (see Sect.4.3.2) characterizing the influence of electrostatic repulsion on the frequency of diffusion collisions are also different.

Spin Exchange Involving Charged Nitroxide Radicals

There are some data in the literature [4.17] on the broadening of ESR lines of charged nitroxide radicals as a result of their spin exchange with each other as well as with potassium ferricyanide anions in $K_3Fe(CN)_6$ solution and with dibezenechromium cations in $Cr(C_6H_6)_2I$ solution. The most interesting results are presented in Tables 4.21-23.

Table 4.21 shows the spin exchange between nitroxide radicals and ferricyanide anions to be sensitive to electrostatic interactions. Indeed as one passes from negatively charged radicals to neutral and further to positively charged ones the constant K_e increases. A 0.06 M phosphatic buffer solution was used as the medium in the above experiments [4.17]. Under these conditions the bulk of ferricyanide seems to exist as an ion pair $KFe(CN)_6^{2-}$ rather than a free $Fe(CN)_6^{3-}$ anion as was believed [4.17]. Indeed, already within the

Table 4.21. Values of $K_e = fK_D$ for spin exchange between radicals and $KFe(CN)_6^{2-}$ in aqueous solution [4.17]

Radical		$K_e \cdot 10^{-9}$ [l/mol · s]
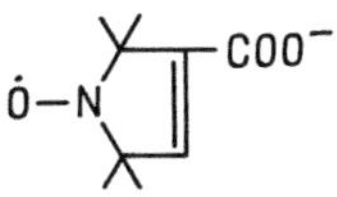	I	0.18
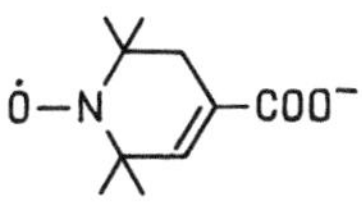	II	0.23
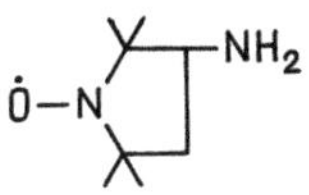	III	0.91
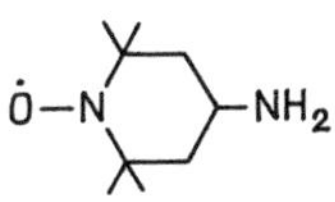	IV	0.98
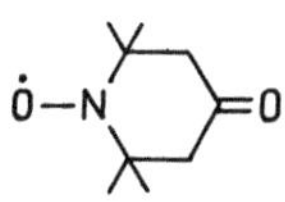	V	1.22
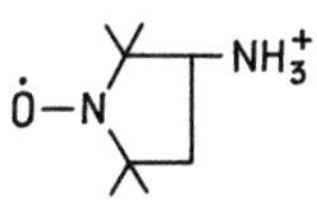	VI	1.68
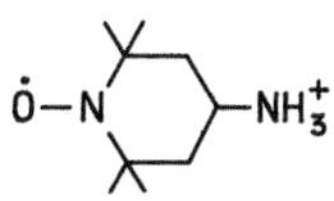	VII	3.50

range of ionic force values $\mu = 0.01 - 0.04$ (smaller than for 0.06 M of phosphate buffer solution) the effective charge of the ferrocyanide anions is 2 rather than 3 [4.64]. On the other hand, the binding of the second cation K^+ in 0.06 M solution does not seem to be completed yet. Under these circumstances the observed [4.17] linearity (see Fig.4.20) of the dependences of ΔH for radical VII (see Table 4.21) on the concentration of $K_3Fe(CN)_6$ up to $C = 1.3 \cdot 10^{-2}$ mol/l (i.e., to $C_{K^+} = 0.06 + 3 \cdot 0.013 \approx 0.1$ mol/l) testifies to the absence of appreciable association of $KFe(CN)_6^{2-}$ with the second cation K^+.

Table 4.22. Values of K_e and fK_D for spin exchange between free radicals in aqueous solution

Radical		$K_e \cdot 10^{-9}$ [l/mol · s]	$fK_D \cdot 10^{-9}$ [l/mol · s]
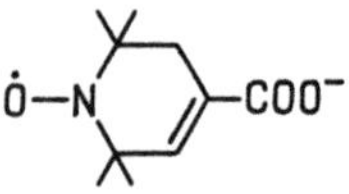	II	0.9	1.8
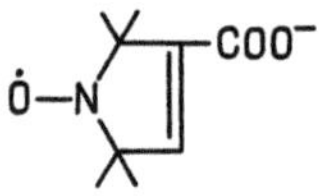	I	1.4	2.8
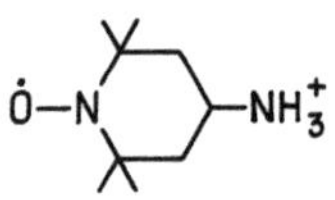	VII	1.6	3.2
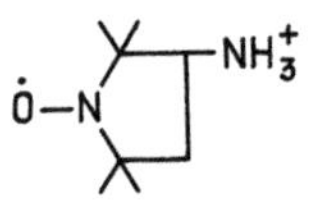	VI	1.9	3.8
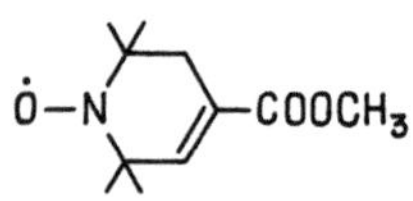	VIII	2.0	4.0
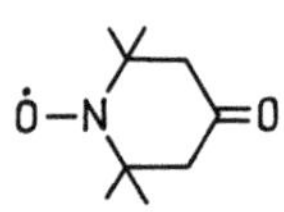	V	2.4	4.8

By using the viscosity criterion in the case of radical V the exchange of nitroxide radicals with ferricyanide anions was shown to be strong. Hence, the values of K_e can be used to quantitatively study the influence of electrostatic interaction on the frequency of collisions of nitroxide radicals with $KFe(CN)_6^{2-}$ anions.

Since the $KFe(CN)_6^{2-}$ anion has a short T_1, the constant K_e coincides with the value of fK_D for strong spin exchange with this anion.

Table 4.22 summarizes the values of K_e calculated from the data available in the literature [4.17] for the processes of spin exchange between charged nitroxide radicals together with typical values of K_e for the exchange between neutral radicals. As was shown in Sect.4.11, in this case the exchange is also strong, and from the values of K_e one can calculate $fK_D = 2K_e$ whose values are also listed in Table 4.22. It is seen that as one passes from

Table 4.23. Values of K_e and fK_D for spin exchange between charged free radicals and $Cr(C_6H_6)_2^+$ cations [4.17]

Radical		$K_e \cdot 10^{-9}$ [l/mol · s]	$fK_D \cdot 10^{-9}$ [l/mol · s]
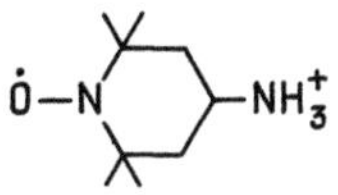	VII	2.0	4.0
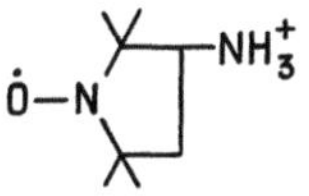	VI	2.0	4.0
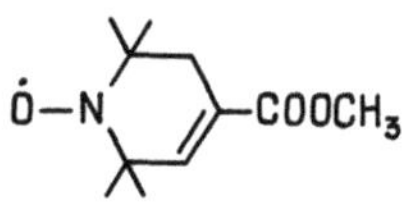	VIII	2.5	5.0
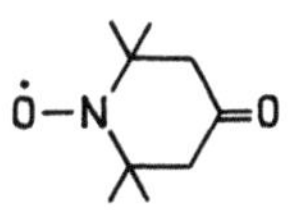	V	2.8	5.6
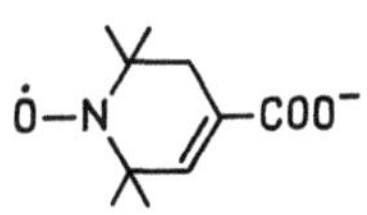	II	2.9	5.8
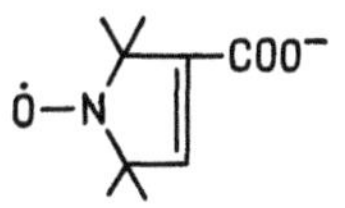	I	3.1	6.2

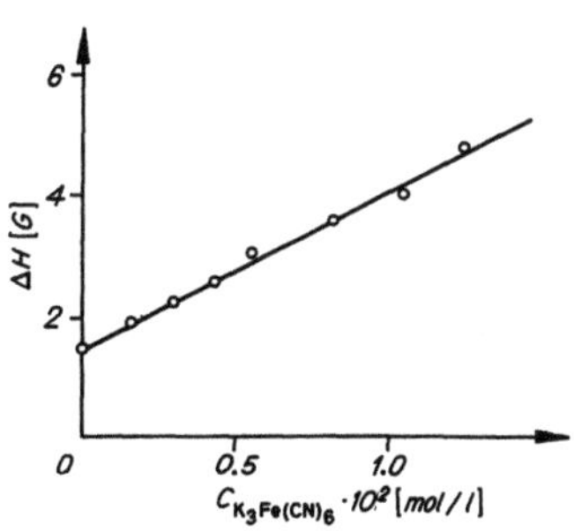

Fig. 4.20. The dependence of radical VII ESR line broadening on the concentration of $K_3Fe(CN)_6$ in aqueous solution [4.17]

electrically neutral to charged radicals the exchange rate constants tend to decrease though very slightly. Even still smaller is the manifestation of electrostatic interaction in spin exchange between charged nitroxide radicals and $Cr(C_6H_6)_2^+$ cations (see Table 4.23).

A comparison was drawn [4.58] between the ESR line broadening of positively charged and electrically neutral nitroxide radicals of a similar structure under the influence of spin exchange with aquo complexes of copper, manganese, and chromium, as well as with the $Cu(en)_2^{2+}$ complex in aqueous solutions with and without electrolyte ($NaNO_3$) added (Table 4.24). For these processes, the electrostatic repulsion of identically charged particles is seen to manifest itself in a systematic, though small decrease of the exchange rate constant in the absence of screening electrolyte. For the $Cr(H_2O)_6^{3+}$ cation association is possible with a counterion in these experiments, just as was the case for $Fe(CN)_6^{3-}$ discussed above.

Table 4.24. Values of K_e and fK_D for spin exchange of charged or uncharged free radicals with charged metal complexes [4.58]

Complex	Radical	$K_e \cdot 10^{-9}$ [l/mol · s]	$fK_D \cdot 10^{-9}$ [l/mol · s]
$Cu(H_2O)_6^{2+}$	R	2.9	5.8
	R in the presence of 2 M of $NaNO_3$	2.4	4.8
	$R'H^+$	1.2	2.4
	$R'H^+$ in the presence of 2 M of $NaNO_3$	2.1	4.2
$Cu(en)_2(H_2O)_2^{2+}$	R	2.9	5.8
	R'	2.4	4.8
	$R'H^+$	1.2	2.4
$Mn(H_2O)_6^{2+}$	R	2.0	3.1
	$R'H^+$	1.1	1.7
$Cr(H_2O)_6^{3+}$	R	1.2	1.9
	$R'H^+$	0.65	1.0

Notation: en = $NH_2CH_2CH_2NH_2$

R = 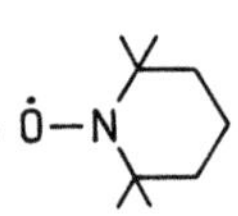 R′ = 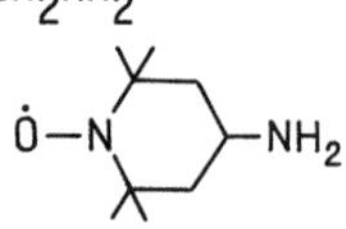 R′H⁺ = 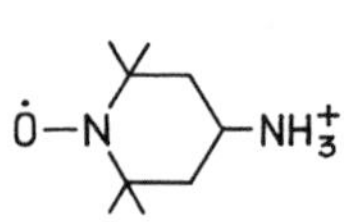 .

Spin Exchange Between Anion Radicals in Organic Solvents

The exchange broadening of ESR lines of anion radicals in organic liquids with small dielectric constants has been studied in detail [4.9,10,60-62] (Table 4.25). For $TCNE^-$ and NPH^- anions a direct proof was obtained of the fulfillment of the strong exchange condition from the viscosity dependences. As follows from Table 4.25 electrostatic interactions do not affect K_e for the exchange between anion radicals in organic solvents with small dielectric constants ε. Indicative of this are the great values of K_e for tetracyano-ethylene anion radicals in dimenthoxyethane and in tetrahydrofuran which are close to the values of K_e for the uncharged DTBN radical.

Theoretical estimates according to (4.23) (see below) show that because of the small values of ε for dimethoxyethane (DME) and tetrahydrofurane (THF) the collision frequency for identically charged particles with $Z_AZ_B = 1$ and $\sigma = 6$ Å must be at least three orders of magnitude smaller than that for electrically neutral particles of the same size. In fact there is only a slight difference between the values of fK_D calculated from K_e, and the theoretically calculated constants $K_D = 8RT/3000\eta$. This testifies to the association of anion radicals with counterions [4.9,10,61]. Direct measurement of the association constant $K \approx 10^6$ l/mol of $TCNE^-$ with K^+ in DME confirms this conclusion [4.61]

The decrease in the steric factor f for $TCNE^-$ in the sequence of counter-ions and solvents $f(K^+;\ THF) > f(K^+;\ DME) > f(DBC,\ K^+;\ DME)$ can be attributed to the screening effect of the solvent and counterion. Indeed, due to the more pronounced solvating ability of DME as compared to THF the $TCNE^-$ anion is probably more screened by DME than by THF molecules. For the same reason the K^+ ion can be solvated in DME more strongly hence its size and screening ability can increase too. The latter increases still more when K^+ complexes with macrocyclic polyether dibenzo-18-crown-6.

One should note a marked (especially when the counterion is Li^+) decrease of factor f when passing from DME and THF to acetonitrile, a solvent with a higher dielectric constant. It is not excluded that in this solvent the ionic pairs partially dissociate [4.65]. Therefore factor f calculated formally from the relation $f = (K_e/p_{max}) \times (3000\eta/8RT)$ for acetonitrile solutions appears to characterize in reality probably not only the steric hinderance to exchange, but also the extent of association of $TCNE^-$ with counterions.

Table 4.25. Rate constants K_e, fK_D, $K_D = 8\ RT/3000\ \eta$, and steric factor f for the spin exchange between charged particles in organic solvents (t = 15°C)

Radical	Counterion	Solvent	$K_e \cdot 10^{-9}$ [l/mol · s]	$fK_D \cdot 10^{-9}$ [l/mol · s]	$K_D \cdot 10^{-9}$ [l/mol · s]	f	References
$TCNE^-$	K^+	DME	4.1 ± 0.6	8.2	12.8	0.64	[4.9]
$TCNE^-$	K^+	THF	6.9 ± 1	13.8	12.6	≈1	[4.9]
$TCNE^-$	(DBC, K^+)	DME	1.5 ± 0.3	3.0	12.8	0.23	[4.61]
DSQ^-	TBA^+	DME	2.9	5.8	12.8	0.45	[4.10]
NPH^- (20°C)	K^+	THF	2.8	5.6	12.6	0.44	[4.60]
DTBN	-	DME	5.5	11	12.8	0.86	[4.10]
DTBN	-	THF	5.0	10	12.6	0.79	[4.10]
$TCNE^-$	Na^+, K^+	AN	3.06 ± 0.07	6.1	19	0.32	[4.62]
$TCNE^-$	Li^+	AN	1.4 ± 0.2	2.8	19	0.15	[4.62]

Notation:

TCNE tetracyanoethylene
DSQ durosemiquinone
NPH naphthalene
TBA tert-butyl ammonium
DTBN di-tert-butyl nitroxide
(DBC, K^+) complex of macrocyclic ester, dibenzo-18-crown-6 with K^+
DME dimethoxy ethane
THF tetrahydrofuran
AN acetonitrile

4.3.2 Comparison of Experiments on Spin Exchange with the Diffusion Theory of Collisions of Charged Particles in Liquid

Compare the data presented in the previous section with the theory of diffusion collisions of charged particles in solution developed by DEBYE in 1942 [4.66].

According to this theory the diffusion rate constant for collisions of spherical ions in very dilute solutions depends on their charges Z_Ae and Z_Be, on the sum of their radii $R = R_A + R_B$, on the diffusion coefficients D_A and D_B, on the macroscopic dielectric constant ε, and on the temperature T

$$K_D(\text{ion}) = (4\pi Z_A Z_B e^2/\varepsilon kT)(D_A + D_B) \times [\exp(Z_A Z_B e^2/\varepsilon RkT)-1]^{-1} \quad . \tag{4.23}$$

Comparison of (4.1) with (4.23) shows the ratio of the constants K_D for charged and for neutral spherical particles of the same size to be equal to

$$Q = K_D(\text{ion})/K_D = y/[\exp(y)-1] \ , \tag{4.24}$$

where

$$y = Z_A Z_B e^2/\varepsilon RkT \quad . \tag{4.25}$$

It is of interest to note that for charged particles the collision time is also changed along with K_D in accord with the expression

$$\tau_c(\mathrm{ion})/\tau_c = [1 - \exp(-y)]/y \quad . \qquad (4.26)$$

From (4.23,24) it follows that for strong Coulomb interaction when $Z_A Z_B e^2/\varepsilon R \gg kT$ the electrostatic repulsion must lead to a considerable decrease in K_D for a collision of identically charged particles, while the electrostatic attraction leads only to a relatively small increase in K_D. And, vice versa, the duration of a collision must greatly increase in the case of attraction and slightly decrease in the case of repulsion [see (4.26)].

Equations (4.23-26) were obtained for the diffusion of charged spherical particles in continuous media. Therefore they are strictly speaking, only valid for a situation when the particles have shapes close to spherical and when their size is much larger than that of the solvent molecules. Experiments on spin exchange make it possible to experimentally verify the applicability of these formulae in describing collisions of real particles of molecular sizes. In order to do this one can, e.g., make use of the data on the dependence of exchange broadening on the concentration of diamagnetic electrolytes (see Fig.4.19). Indeed, the ratio of the values of K_e at low concentrations of electrolyte in solution ($K_e^{(0)}$) to the extreme values of K_e at its high concentrations ($K_e^{(\infty)}$), when the influence of electrostatic interactions is completely neutralized, may be regarded as an experimental value of Q

$$K_e^{(0)}/K_e^{(\infty)} = Q_{\mathrm{exper}} \quad . \qquad (4.27)$$

In this case the ratio $K_e^{(0)}/K_e^{(\infty)}$ is equal to that of the corresponding values of the exchange broadenings $\Delta H_e^{(0)}/C$ and $\Delta H_e^{(\infty)}/C$.

Note that this way of determining Q_{exper} automatically excludes the above-mentioned possible errors [4.54] in measuring the absolute values of $K_e^{(0)}$ and $K_e^{(\infty)}$ for the exchange $(SO_3)_2NO^{2-} + (SO_3)_2NO^{2-}$.

For the exchange of charged nitroxide radicals with one another, with a $KFe(CN)_6^{2-}$ anion and with a $Cr(C_6H_6)_2^+$ cation one can take the value of Q_{exper} to equal the ratio of the average values of K_e for all the charged and electrically neutral radicals studied, e.g.,

$$Q(R^+) = K_e(R^\pm + R^\pm)/K_e(R + R) \quad , \qquad (4.28)$$

$$Q(R^\pm,\ Cr(C_6H_6)_2^+) = K_e[R^\pm + Cr(C_6H_6)_2^+]/K_e[R + Cr(C_6H_6)_2^+] \quad . \qquad (4.29)$$

For the processes of exchange of the $R'H^+$ radical (see Table 4.24) with positively charged complexes the value of Q_{exper} was chosen to be the ratio of the constant K_e for these processes to the average value $\bar{K}_e$ for the exchange of a given complex with electrically neutral radicals

$$Q_{exper} = K_e(R'H^+ + M^{2+})/\bar{K}_e(R + M^{2+}) \quad . \tag{4.30}$$

Here for all M^{2+} the same values of Q_{exper} were obtained. For the exchange of $R'H^+$ with $Cu(H_2O)_6^{2+}$ the value of Q_{exper} was also calculated from the ratio of the values of K_e without $NaNO_3$ additions and with 2M of $NaNO_3$.

The values of Q_{exper} obtained by the above techniques are compared in Table 4.26 with those calculated by using (4.24,25). In the calculations ε was assumed to be equal to 78, and the average radii of $(SO_3)_2NO^{2-}$, R^+, R^-, $Cr(C_6H_6)_2^+$ and of doubly charged complexes of copper and manganese as well as of ferricyanide anions to 3.5 Å [4.10,58]. From Table 4.26 it is seen that the trends and scale of variation of the experimental values of Q coincide with theoretical predictions. Thus, electrostatic repulsion decreases Q in a much greater degree than electrostatic attraction increases it (cf the value of Q for systems 1 and 8 with the same value of $|Z_A \cdot Z_B| = 4$). In all cases the discrepancy between Q_{theor} and Q_{exper} does not exceed 50%.

Taking into account the assumptions made both in obtaining the theoretical formulae and in processing the experimental spin exchange data such an agreement should be regarded as quite satisfactory.

However, in many cases the application of the Debye formula encounters some difficulties. Particularly unreliable are the theoretical estimates of the influence of electrostatic interactions on K_D in cases when direct association of colliding ions with counterions is possible. We have already met with this kind of association, e.g., in solutions of potassium ferricyanide for which a fraction of anions are bound into the pairs $K^+Fe(CN)_6^{3-}$ so that their total charge is equal to 2 rather than 3. The situation becomes especially complicated in solutions with a large concentration of electrolytes where the effects of charge screening are superimposed on those of ionic association. Great complications also arise in nonaqueous media where the effects of ionic association and charge screening have been studied rather little.

Spin exchange allows one to measure the diffusion rate constants fK_D for collisions of ions under any conditions, including those in which the theoretical estimates of electrostatic effects are practically impossible. The diffusion constants determined by such a method automatically reflect the influence on fK_D of the charge distribution in ions as well as of such factors as association with counterions, charge screening by ionic atmosphere, etc.

Table 4.26. Comparison of theoretical and experimental values of $Q = K_D\ (\text{ion})/K_D\ (\text{neutr.})$

No.	System	$Z_A \cdot Z_B$	Q_{theor}	Q_{exper}
1	$(SO_3)_2NO^{2-} + Co(H_2O)_6^{2+}$	-4	4.2	3.8 [a]
2	$R^+ + KFe(CN)_6^{2-}$	-2	2.3	2.6 [b]
3	$R^- + Cr(C_6H_5)_2^+$	-1	1.6	1.1 [b]
4	$R^+ + R^+,\quad R^- + R^-$	+1	0.6	0.6 [b]
5	$R^+ + Cr(C_6H_5)_2^+$	+1	0.6	0.8 [b]
6	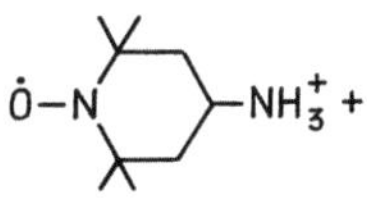 $+M^{2+}$	+2	0.3	0.6 [a] 0.5 [b]
7	$R^- + KFe(CN)_6^{2-}$	+2	0.3	0.2 [b]
8	$(SO_3)_2NO^{2-} + (SO_3)_2NO^{2-}$	+4	0.07	0.10 [a]

[a] Calculated from the ratio of rate constants K_e without and with electrolyte.

[b] Calculated from the ratio of rate constants K_e for spin exchange between charged and neutral particles of similar structure.

4.3.3 Study of Collisions Between Aquo Complexes of Transition Metals

Role of Counterions in Collisions of Aquo Complexes of Vanadyl

As an example demonstrating the possibilities of spin exchange in quantitative studies of the kinetics and mechanism of a collision of charged particles, consider the study of collisions of aquo complexes of vanadyl $VO(H_2O)_5^{2+}$ in concentrated (up to several moles per litre) aqueous solutions of various electrolytes. This system is rather complex for theoretical analysis.

The kinetics of spin exchange between $VO(H_2O)_5^{2+}$ cations in aqueous solutions of vanadyl chloride, nitrate, and perchlorate was found to follow the third- rather than the second-order law as is usually the case [4.56,58,67].

As an example, Figs.4.21-24 present the concentration dependences of the width and the shift δH of ESR lines for solutions of vanadyl salts. The spin exchange rate constants were measured for the solutions of $VO(NO_3)_2$ and $VO(ClO_4)_2$, from the exchange broadening of the hfs components of the ESR spectrum within the range of concentrations 0.035-0.35 mol/l and 0.06-0.64

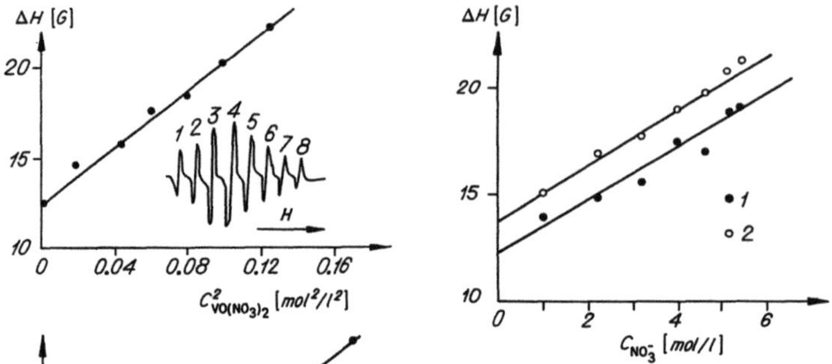

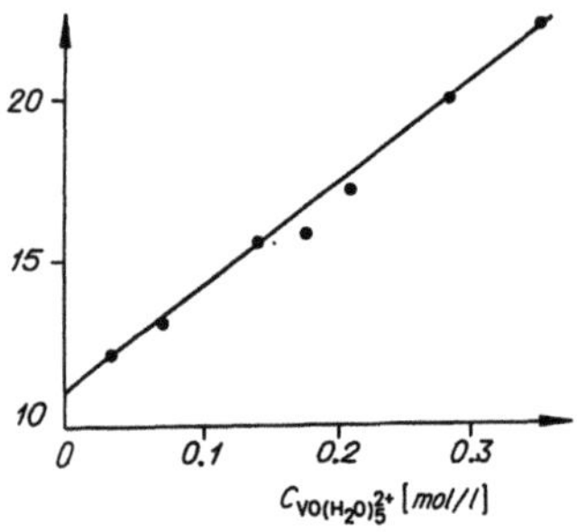

Fig. 4.21. The dependences of the width of a hfs component for $VO(NO_3)_2$ in aqueous solutions on the total concentration of $VO(NO_3)_2$, on the concentration of $VO(H_2O)_5^{2+}$ cations for $C_{NO_3^-}$ = 2.6 mol/l, on that of NO_3^- anions for $C_{VO(H_2O)_5^{2+}}$ = 0.14 mol/l [4.67]. 1: component 4, 2: component 2

mol/l, respectively, while for the solutions of $VOCl_2$, from the exchange broadening of the ESR lines in the range of concentrations 0.03 - 0.66 mol/l and the exchange shift in the range of concentrations 0.66 - 2.2 mol/l. Calculations of the exchange rate constants from btoadening and shift yielded coincident results. Note that the rate constants correspond to termolecular rather than bimolecular collisions in these cases. In all the three systems studied the exchange rate ω_e was found to depend not only on the concentration of paramagnetic aquo complexes of vanadyl but also on that of diamagnetic anions, A^-

$$\omega_e = K_e' C_{VO(H_2O)_5^{2+}} C_{A^-} , \tag{4.31}$$

where $A^- = Cl^-$, NO_3^-, or ClO_4^- (see, e.g., concentration dependences in Fig. 4.21). The concentration of anions was varied by changing that of vanadyl salts as well as by adding HCl, LiCl, HNO_3, $HClO_4$, $LiClO_4$, and $Ba(ClO_4)_2$. Experimental values of rate constants are independent of the nature and concentration of diamagnetic cations.

The observed dependence of ω_e on $C_{VO(H_2O)_5^{2+}}$ and C_{A^-} can be accounted for by assuming the spin exchange in the systems studied to occur during termolecular collisions

$$\uparrow VO(H_2O)_5^{2+} + A^- + \downarrow VO(H_2O)_5^{2+}$$
$$\rightarrow \downarrow VO(H_2O)_5^{2+} + A^- + \uparrow VO(H_2O)_5^{2+} , \tag{4.32}$$

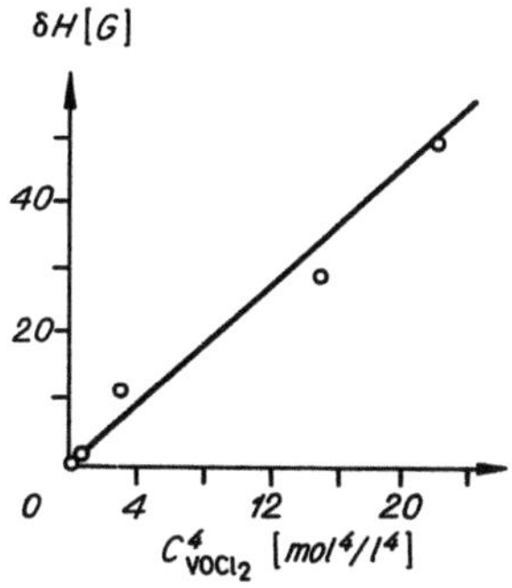

Fig. 4.22. Concentration dependence of the hf component's shift of $VOCl_2$ in aqueous solutions [4.67]

where the directions of the arrows correspond to the +1/2 and -1/2 projections of the spin of the unpaired electron of $VO(H_2O)_5^{2+}$ onto the direction of the external mangetic field.

Indeed, the measured exchange rate constants are of the same order of magnitude as termolecular diffusion rate constants in liquid.

The results obtained, however, can also be interpreted within the framework of a bimolecular mechanism by assuming that the spin exchange involves complexes $VO(H_2O)_5^{2+} \cdot A^-$

$$\uparrow VO(H_2O)_5^{2+} + \downarrow VO(H_2O)_5^{2+} \cdot A^-$$

$$\rightarrow \downarrow VO(H_2O)_5^{2+} + \uparrow VO(H_2O)_5^{2+} \cdot A^- \tag{4.33}$$

which account for but a small fraction of the total amount of vanadyl in solution. Indeed, in this case

$$C_{VO(H_2O)_5^{2+} \circ A^-} = KC_{VO(H_2O)_5^{2+}} \cdot C_{A^-} \quad . \tag{4.34}$$

Consequently,

$$\omega_e = K_e C_{VO(H_2O)_5^{2+} \cdot A^-} = K_e K C_{VO(H_2O)_5^{2+}} \cdot C_{A^-} \tag{4.35}$$

and

$$K_e' = K_e K \quad . \tag{4.36}$$

As can be seen from Table 4.27 the spin exchange rate constants in the solutions of $VOCl_2$, $VOClO_4$, and $VO(NO_3)_2$ are practically the same. This seems to mean that in the triple complex $\downarrow VO(H_2O)_5^{2+} \cdot A^- \circ \uparrow VO(H_2O)_5^{2+}$ the anion A^- and the $VO(H_2O)_5^{2+}$ cations are only bound by electrostatic forces, and the role of the anions in the exchange is reduced to decreasing the Coulomb repulsion between the cations of vanadyl. The $VO(H_2O)_5^{2+} \cdot A^-$ complex is an ionic pair while the mechanism (4.32) is a particular case of the mechanism (4.33) when the lifetime of the $VO(H_2O)_5^{2+} \circ A^-$ complex is equal to the duration of a collision between the ions $VO(H_2O)_5^{2+}$ and A^-.

In contrast to $VOCl_2$, $VO(NO_3)_2$, and $VO(ClO_4)_2$ the dependence of ω_e on salt concentration in the case of $VOSO_4$ is linear (Fig.4.23). It seems to be due to the high stability of complexes of vanadyl with SO_4^{2-} (stability constant $K = 3 \cdot 10^2$ mol/l) as distinct from the low stability of complexes containing Cl^-, NO_3^-, and ClO_4^- [4.67]. Thus, in the region of concentration studied most of the vanadyl ions exist as neutral particles of $VO(H_2O)_nSO_4$. The concentration dependence of the ESR linewidth is also linear in the case of another uncharged vanadyl complex; $VO(acac)_2$ (see Fig.3.3).

The data obtained show that diamagnetic anions can play an important role in decreasing the electrostatic repulsion for collisions of vanadyl cations. At sufficiently high concentrations the termolecular collision of

Table 4.27. Rate constants of spin exchange in solutions of vanadyl salts

Compound	$K_e' \cdot 10^{-8}$ [$l^2/mol^2 \cdot s$]
$VOCl_2$	
1) exchange broadening	1.4 ± 0.2
2) exchange shift	2 ± 1
$VO(NO_3)_2$	1.6 ± 0.2
$VO(ClO_4)_2$	1.1 ± 0.1

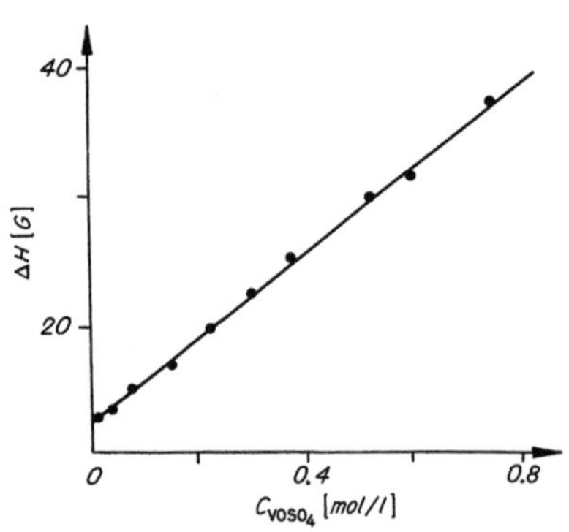

Fig. 4.23. The concentration broadening of $VOSO_4$ ESR line in aqueous solution [4.67]

two vanadyl cations and of an anion proves to be a much faster process than their ordinary binary collision, i.e., the collisions of charged particles are catalyzed by counterions.

An Example of the Process with Weak Influence of Electrostatic Repulsion

When discussing the spin exchange between anion radicals in organic solutions with small dielectric constant we encountered a situation where the electrostatic interactions did not practically manifest themselves in the spin exchange. This was caused by the formation of electrically neutral complexes of anion radicals with counterions. At the same time for aqueous solutions with large dielectric constant where the association with counterions is not so strong the influence of electrostatic interactions proved to be essential in all cases.

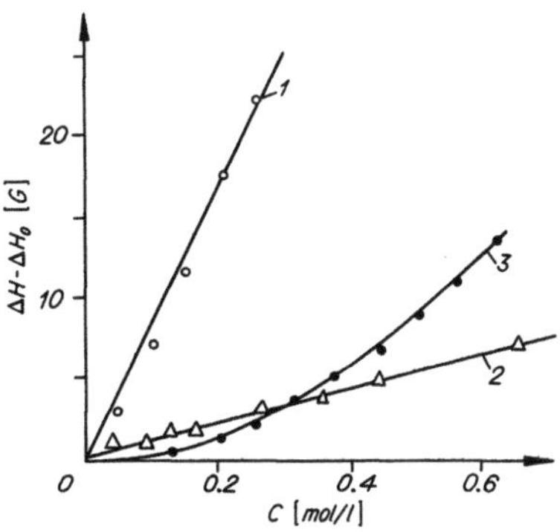

Fig. 4.24. The concentration broadening of the ESR lines of $VO(H_2O)_5^{2+}$ (1) and $Mn(H_2O)_6^{2+}$ (2) in the presence of copper nitrate and of $VO(H_2O)_5^{2+}$ in solutions of vanadyl perchlorate (3) [4.57]

The only exception to this rule so far known is the spin exchange of copper aquo complex $Cu(H_2O)_6^{2+}$ with those of manganese, $Mn(H_2O)_6^{2+}$, and of vanadyl, $VO(H_2O)_5^{2+}$ [4.57] (Fig.4.24). The paramagnetic ions were introduced into solutions as the salts $Mn(NO_3)_2$, $VO(ClO_4)_2$, and $Cu(NO_3)_2$. Wherever possible the constants of broadening were measured from the ESR spectra of both partners. When the complexes of manganese, vanadyl, and copper were used as indicators their concentrations were 0.05, 0.1, and 0.2 mol/l, respectively, and when used as broadening agents their concentrations varied from 0.01 to a few tenths of a mol/l.

The broadening constants K_b for the systems studied are listed in Table 4.28.

According to the criteria formulated in Sect.3.3 for the pairs $Cu(H_2O)_6^{2+}$ + $Mn(H_2O)_6^{2+}$ and $Cu(H_2O)_6^{2+}$ + $VO(H_2O)_6^{2+}$ the dipole-dipole contribution to K_b may be neglected and one may assume $K_b = K_e$. Note that in accord with theory (see Sect.2.5.5) for the system $Cu(H_2O)_6^{2+} + Mn(H_2O)_6^{2+}$ the rate constants determined from the line broadening of $Mn(H_2O)_6^{2+}$ and $Cu(H_2O)_6^{2+}$ differ from each other by about one order of magnitude because of the difference in the factor P_{max}.

Table 4.28. Broadening constants, K_b (l/mol · s), for interaction between aquo complexes

Broadening agent	Indicator $Mn(H_2O)_6^{2+}$	$VO(H_2O)_5^{2+}$	$Cu(H_2O)_6^{2+}$
$Mn(H_2O)_6^{2+}$	$3.3 \cdot 10^8$	$6.2 \cdot 10^8$	$1.5 \cdot 10^9$
$VO(H_2O)_5^{2+}$	$9 \cdot 10^7$	-	-
$Cu(H_2O)_6^{2+}$	$1.6 \cdot 10^8$	$1.3 \cdot 10^9$	-

Table 4.29 summarizes the values of the constant $K_e = K_b$ and the parameters K_e/P_{max} for the systems $Cu(H_2O)_6^{2+} + Mn(H_2O)_6^{2+}$ and $Cu(H_2O)_6^{2+} + VO(H_2O)_5^{2+}$. The values of P_{max} were taken from Table 4.1. For comparison the table also includes the data on the spin exchange of aquo complexes with the uncharged TANONE radical, and on the spin exchange of uncharged acetylacetonates of copper, manganese, and vanadyl with different partners in organic solvents. The constants K_e and the parameters K_e/P_{max} were recalculated (normalized) in this case to fit the viscosity $\eta = 1$ cP using the relations $K_e = \eta \cdot (K_e)_{exper}$ and $K_e/P_{max} = \eta(K_e)_{exper}/P_{max}$ where η is the viscosity of the solution in which the rate constant $(K_e)_{exper}$ was determined experimentally.

Table 4.29 shows that for the exchange between the charged particles $Cu(H_2O)_6^{2+} + Mn(H_2O)_6^{2+}$ and $Cu(H_2O)_6^{2+} + VO(H_2O)_5^{2+}$ the values of K_e/P_{max} turn out to be of the same order of magnitude as for the exchange involving uncharged particles. Thus, in the spin exchange of an aquo complex of copper with complexes of manganese and vanadyl, electrostatic repulsion does not manifest itself.

To explain the lack of influence of electrostatic repulsion on the exchange processes involving $Cu(H_2O)_6^{2+}$ two hypotheses have been suggested: 1) unusually strong association of $Cu(H_2O)_6^{2+}$ with counterions in comparison with other aquo complexes; 2) realization of spin exchange at long distances due to delocalization of spin density over the system of hydrogen bonds of water molecules.

The currently available data do not allow one to give unambiguous preference to any of these explanations.

Table 4.29. Constants K_e and parameters K_e/P_{max} for spin exchange involving aquo complexes and acetylacetonates of copper, manganese, and vanadyl

No.	Indicator	Broadening agent	$K_e \cdot 10^{-9}$ [l/mol · s]	$(K_e/P_{max}) \cdot 10^{-9}$ [l/mol · s]
1	$Cu(H_2O)_6^{2+}$	$Mn(H_2O)_6^{2+}$	1.5	2.3
2	$Mn(H_2O)_6^{2+}$	$Cu(H_2O)_6^{2+}$	0.16	2.9
3	$VO(H_2O)_5^{2+}$	$Cu(H_2O)_6^{2+}$	1.3	2.6
4	TANONE	$Cu(H_2O)_6^{2+}$	1.4	2.8
5	TANONE	$Mn(H_2O)_6^{2+}$	2.0	3.1
6	TANONE	$VO(H_2O)_5^{2+}$	1.2	2.4
7	TANONE	$Cu(acac)_2$	1.3	2.6
8	TANONE	$Mn(acac)_2$	2.9	4.5
9	TANONE	$VO(acac)_2$	0.9	1.8
10	DPPH	$Cu(acac)_2$	1.3	2.6
11	BDC	$Cu(acac)_2$	0.7	1.4
12	BDC	$Mn(acac)_2$	1.6	2.4
13	$VO(acac)_2$	$Cu(acac)_2$	1.2	2.4
14	$VO(acac)_2$	$VO(acac)_2$	0.8	1.6

Notation: TANONE - 2,2,6,6-tetramethyl-4-oxopiperidine-1-oxyl; DPPH - αα-diphenyl-β-picryl-hydrazyl; BDC - bisdiphenylchromium cation; acac - acetylacetonate anion.

Note: Broadening constants K_b used in calculations of K_e were taken from [4.57,58] for systems 1-3, 13, 14; from [4.27,58] for 4-6; and from [4.45] for 7-12.

4.4 Biological Applications of Spin Exchange

Application of spin exchange in molecular biology is most often based on the use of a so-called spin probe technique [4.68]. In this technique the chemically inert paramagnetic particles (spin probes) are introduced into biological systems and then the kinetics of the spin exchange of these probes is studied with one another or with the paramagnetic centers inherent to the biological objects or introduced artificially into them (spin labels). Examples of utilization of this method are rather numerous [4.68]. We shall therefore restrict ourselves to a brief consideration of the range of problems that can be solved using the spin probe technique.

Paramagnetics Used as Spin Probes

Stable free nitroxide radicals, potassium ferricyanide, and dibenzenechromium iodide are most often used as spin probes. These compounds are chemically inert enough and dissolve readily in water. Many nitroxide radicals are also easily soluble in organic media which allows one to introduce them as probes into hydrophobic regions of biological structures.

4.4.1 Measurement of Binary Collision Rate of Molecules in Biological Systems

To measure the rate of binary collisions in biological systems, just as in nonbiological systems, one should make use of processes with strong spin exchange between probes, that is, processes for which conditions $J\tau_c \gg 1$ or $J(T_1\tau_c)^{1/2} \gg 1$ are fulfilled. An example of a detailed investigation in this field is the estimation of the rate of binary collisions of molecules in two-dimensional diffusion motion in biological membranes [4.69-72].

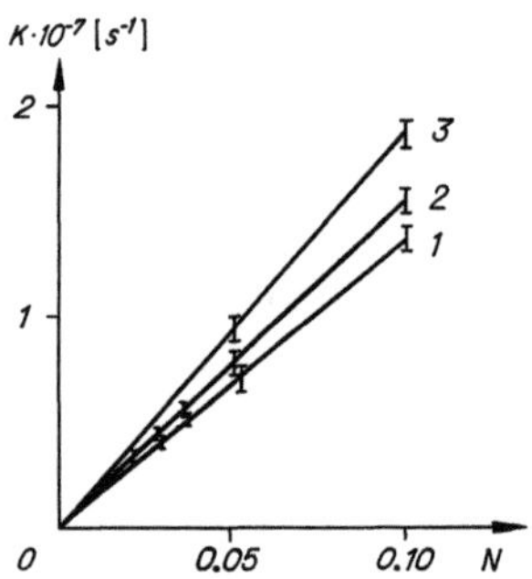

Fig. 4.25. The dependences of the binary collision rates for radical fragments of spin-labelled lecithin molecules on their mole fraction in membranes at the temperatures of 50°C (1), 60°C (2), and 70°C (3) [4.72]

Figure 4.25 depicts the concentration dependences of the rate of collisions between radical fragments of spin-labelled lecithin phospho-lipid molecules

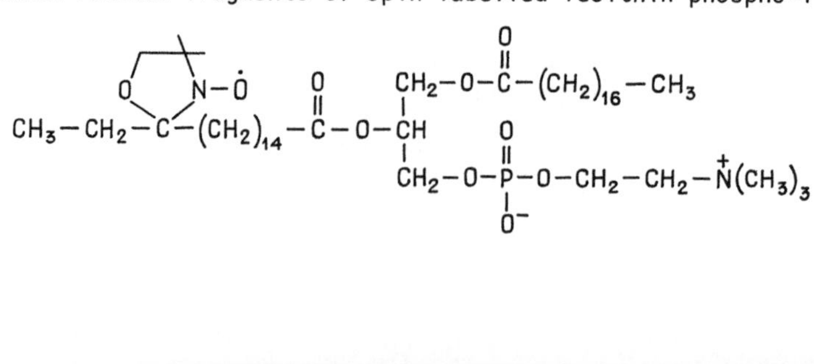

in membranes built of egg lecithin and cholesterol in the ratio of 4:1. The rate (the number of collisions K of one paramagnetic center with all the others per unit of time) was calculated under the assumption of strong exchange from the spin exchange rate ω_e

$$K = \omega_e/(1 - \varphi)p_{max} = 3\omega_e \quad , \tag{4.37}$$

where $\varphi = 1/3$; $P_{max} = 1/2$. The spin exchange rate ω_e was measured in the region of the exchange broadening of the ESR line for spin-labelled lecithin by comparing the experimentally observed ESR spectra with the theoretical spectra calculated for different values of ω_e by means of (2.77). The predominance of the exchange contribution to the concentration broadening of ESR lines over the dipole-dipole contribution was controlled by the viscosity dependence of broadening (see Sect.3.3). The membrane viscosity was varied by varying the temperature (Fig.4.25). These measurements show also that the strong exchange condition for the collisions of radical fragments of spin-labelled lecithin is fulfilled.

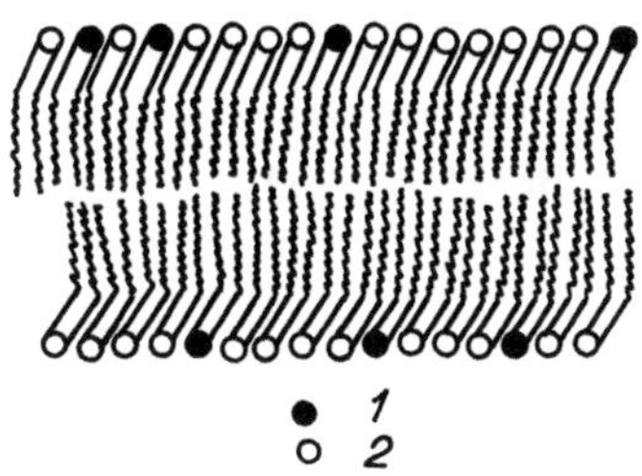

Fig. 4.26. A schematic model of the structure of a spin-labelled bilayer membrane (side view). 1 - The polar groups of spin-labelled molecules; 2 - the polar groups of usual molecules

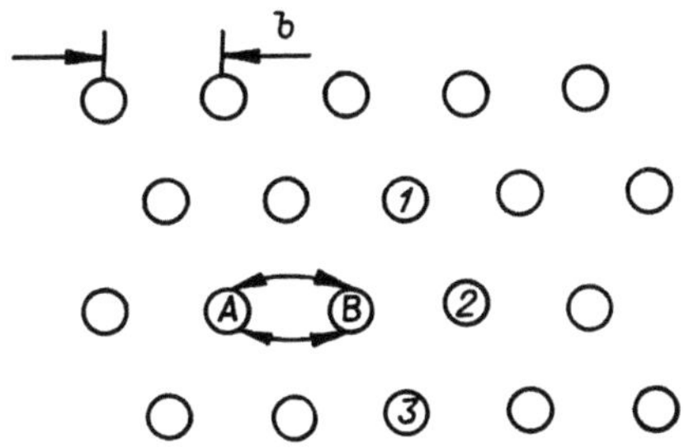

Fig. 4.27. A scheme of the two-dimensional diffusion in the membrane depicted in Fig.4.26 (top view). Molecule A acquires new neighbors 1, 2 and 3 as a result of the position exchange with molecule B; b is the distance between the neighboring molecules

A schematic model of the spin-labelled bilayer membrane is depicted in Fig.4.26 [4.70,72]while Fig.4.27 represents the process of two-dimensional diffusion in such a membrane [4.72]. For simplicity the lipid molecules are approximated by cylinders which are packed in a hexagonal lattice with a characteristic distance b $\cong$ 9 Å [4.70]. The diffusion is assumed to occur via an exchange of position of the neighboring molecules A and B. After the exchange each molecule acquires three new neighbors (molecules 1, 2, 3 for particle A). In such a model the number K of spin-labelled binary collisions is related to the coefficient D of two-dimensional molecular diffusion in the membrane by the equation

$$D = Kb^2/(12N) \quad , \tag{4.38}$$

where N is the fraction of spin-labelled particles in the membrane.

The values of D at different temperatures obtained by means of (4.38) for the membrane under consideration are presented in the Arrhenius coordinates in Fig.4.28. Extrapolation to 25°C yields the value of $D = 7 \cdot 10^{-8}$ cm^2/s while direct measurement of the macroscopic coefficient of two-dimensional spin label diffusion in the same membrane gives $D = 1.8 \cdot 10^{-8}$ cm^2/s. Bearing in mind the simplified character of the diffusion model in Fig.4.27 and the somewhat arbitrary choice of the diffusion jump length b one should recognize this agreement as rather good. Better agreement can be obtained by assuming a shorter length of the diffusion jump or a smaller number of new neighbors acquired as a result of one jump. The diffusion coefficients of spin-labelled molecules close to $D \simeq 10^{-7}$-10^{-8} cm^2/s also follow from other investigations of membranes [4.69,71]. Comparison of these values with typical values of diffusion coefficients $D \simeq 10^{-5}$ cm^2/s for liquids with the viscosity $\eta = 1$ cP shows that the membrane molecules diffuse 10^2 to 10^3 times as slowly.

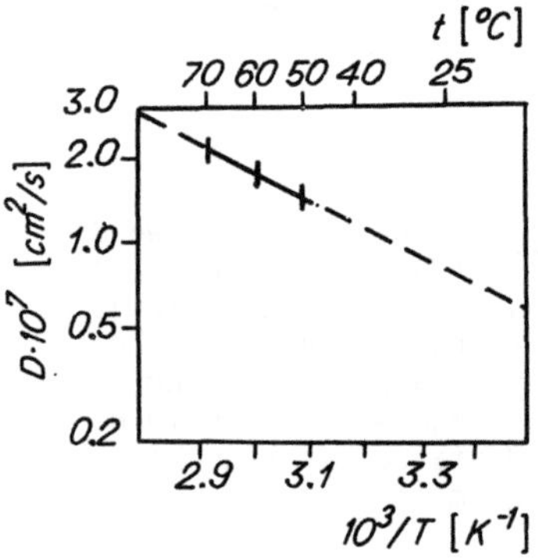

Fig. 4.28. The dependence of the two-dimensional diffusion coefficient on temperature for membranes built of egg lecithin and cholesterol in the ratio of 4:1

Introducing into a biological system two or more closely located and covalently bound spin labels one can study intramolecular collisions by using exactly the same techniques as in the case of nitroxide biradicals (Sect. 4.1.3). Thus, in the ESR spectrum of the spin-labelled UPU molecule (Fig.4.29) broadened components 2 and 4 were observed which are typical for the biradical and testify to spin exchange occurring between nitroxide fragments [4.73].

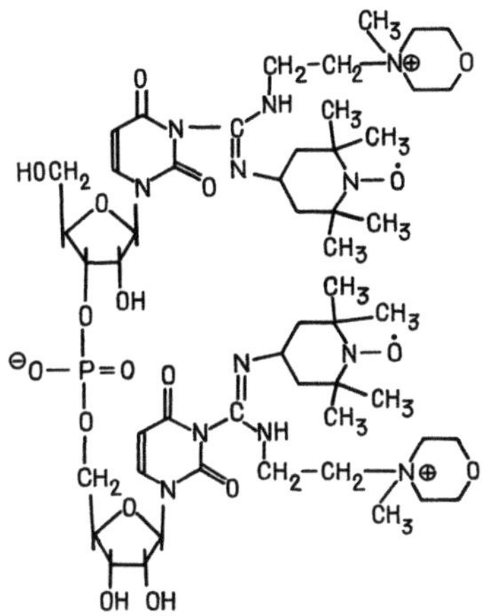

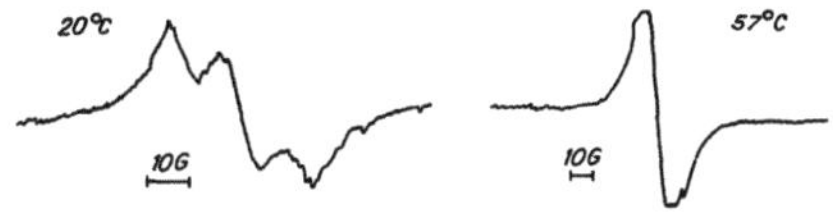

Fig. 4.29. The structure of dinucleotide UPU modified by spin-labelled S-carbodiimide

Fig. 4.30. The ESR spectra of a $2 \cdot 10^{-3}$M solution of a fully modified poly-U [4.73]. Exchange narrowing takes place at higher temperature

From the analysis of the ESR spectrum the time between radical collisions was estimated: $\tau_0 \cong 10^{-9}$ s at 85°C. This result was used [4.73] to discuss the mechanism of spin label collisions in solutions of polyuridylic acid containing the same spin labels, as UPU does, in each nucleotide fragment. As temperature rose, one could observe in such a system the exchange broadening of triplet components of the ESR spectrum, its merging into one line, and subsequent narrowing of this singlet line (Fig.4.30). Such behavior of the spectrum indicates that all paramagnetic fragments are involved in the intramolecular spin exchange. In principle, one can imagine two models of collisions of paramagnetic fragments resulting in this kind of spectrum transformation of spin-labelled poly-U: 1) with collisions possible only between the neighboring labels and 2) with collisions possible between any labels as is the case in solutions. The data available do not yet allow an unambiguous choice to be made between these possibilities. However, irrespective of the details of the models the spin exchange frequency ω_e and hence the time between collisions $\tau_0 = \omega_e^{-1}$ can be estimated from the spectrum of Fig.4.30 with the help of the same equations as for monoradical solutions (see Sect.3.1). Such an estimation gives, e.g., for 85° $\omega_e \cong 7 \cdot 10^8$ rad/s and $\tau_0 \cong 10^{-9}$ s which coincides with the value of τ_0 for a biradical of spin-labelled UPU. Such a coincidence allows us to conclude that spin exchange in the polymer and that in the dinucleotide are induced by the same type of intramolecular motion.

4.4.2 Study of Location of Paramagnetic Centers in Biopolymers

Paramagnetic metals atoms are present in many enzymes and other biopolymers. Study of spin exchange between these paramagnetic atoms and spin probes freely diffusing in aqueous solution makes it possible to answer the question of whether the paramagnetic centers are hidden inside the biopolymer globules or are on their surface. For this purpose one should compare the rate constants of spin exchange of the probe with the paramagnetic centers of the biopolymer and with its low-molecular analogs. As an example, in Table 4.30 a comparison is drawn between the rate constants of spin exchange of a probe, a free radical (TANONE), with ferrihemoglobine and oxidized form of cytochrome C which contain porphyrin complexes of Fe(III) as well as with their low-molecular analogs - pyridinates of Fe(III)hemin and Fe(III) ethyoporphyrin.

Table 4.30. Comparison of the rate constants of spin exchange of the probe (TANONE) with the Fe(III) porphyrins in biopolymers and with their low-molecular analogs [4.68,74]

Fe(III) porphyrin complex	$K_e \cdot 10^{-9}$ [l/mol · s]	Maximum complex concentration · 10^{-2} [mol/l]
Ferrihemoglobin	< 0.1	0.60 [a]
Oxidized cytochrome C	< 0.1	0.60 [a]
Heminpyridinate	2.4	1.64
Ethioporphyrine pyridinate	2.6	1.64

[a] Per hemin unit.

As seen from this table, the rate constants of spin exchange of TANONE with low-molecular complexes are close to the maximum values characteristic of the strong exchange processes involving TANONE (see Sect.4.1.1). However, for Fe(III) complexes with biopolymers these constants are at least 25 times smaller. The result obtained allows a conclusion that in solution the active centers of hemoglobine and cytochrome C are located inside the protein globule and so are unattainable for collisions with the spin probe [4.68].

In a similar way one can also study the location in the biopolymer globules of paramagnetic centers specially introduced into them, as spin labels. Table 4.31 lists the rate constants of spin exchange of a probe (ferricyanide anion) with a spin label - nitroxide free radical (R)

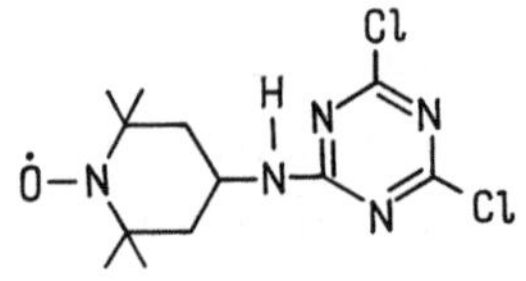

linked to histidine groups of different proteins. The fulfillment of the strong exchange conditions for collisions of a ferricyanide anion with nitroxide radicals was checked [4.17] on an example of the TANONE radical by the viscosity dependences of K_e (see Sect.3.4).

Table 4.31. Rate constants of spin exchange between a ferricyanide anion and a spin label R attached to proteins [4.68]

Spin-labelled protein group	pH	$K_e \cdot 10^{-9}$ [l/mol · s]
His of myoglobin	7.0	0.50
His-15 of lysozyme	6.0	0.43
His, lys of albumin	9.0	0.33
His of silk fibroin	7.0	0.03

As is seen from Table 4.31, the label bound to silk fibroin is far less attainable for collisions with the probe than those bound to other proteins.

The spin exchange technique may also be very useful in studying the location of the active sites of enzymes built in the membrane. Appication of a probe soluble in water but insoluble in the membrane allows one in this case to find out whether the active site of the enzyme is located on the membrane surface or inside the membrane in the hydrophobic region.

Consider as an example the study of liver microsomes which include heme-containing enzyme, cytochrome P-450 [4.75]. In this case use was made of an isocyanide spin label

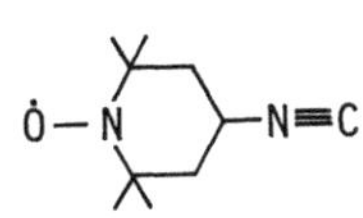

which was supposed to specifically couple with the heme of cytochrome P-450. The interaction of this spin label with ferricyanide ion, the spin probe soluble in water, was studied. The probe proved to influence the spectrum of the microsome label quite insignificantly though it effectively broadened the ESR spectrum lines of the free spin label. Hence it follows that the spin label is located in the hydrophobic pockets rather than on the microsome surface.

Addition of sodium deoxycholate, urea, p-chloromercuribenzoate was shown to make the label more accessible to a ferricyanide ion.

Unfortunately the above results refer probably to an isocyanide label which is located in the lipid phase of the microsome membrane rather than in the active site of the enzyme. Subsequent studies [4.76,77] showed the ESR spectrum of the isocyanide labels bound to the heme of cytochrome P-450 to be unobservable due to large broadening. In contrast to the ESR spectrum of the isocyanide spin label, that of the metyrapone spin label

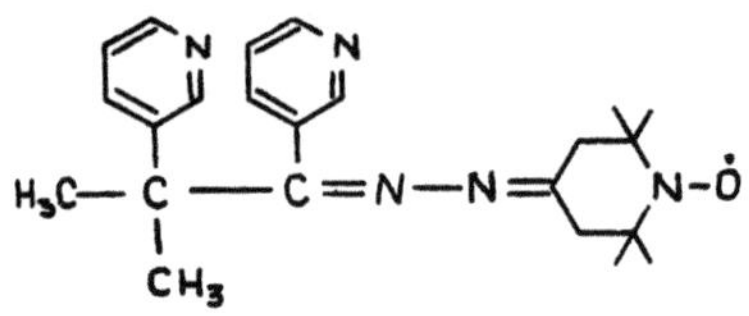

in a complex with the heme of cytochrome P-450 can be readily observed [4.77] which is probably due to a greater distance between the heme iron and the NO group in this case. Thus, this complex might serve as a convenient system for elucidating the accessibility of heme via the spin exchange technique.

Quantitative Interpretation of the Values of K_e for Spin Exchange Processes Involving Paramagnetic Centers of Biopolymers

The above examples show convincingly that the problem of the accessibility of paramagnetic centers in biopolymers for collisions with probes from solutions can be solved rather simply on the qualitative level. However, the quantitative interpretation of the values of K_e observed for spin exchange involving paramagnetic centers of biopolymers is a far more complex problem. Unfortunately research in this field is only just beginning and so far only some general remarks can be made on this problem. If the exchange rates of a probe with the biopolymer paramagnetic center and with the low-molecular analog of this center are nearly equal, then one can reliably state the absence of any significant obstacles to collisions of the probe with this center. If the probe is electrically neutral it simply indicates the absence of steric hindrance. If the probe has an electric charge, one can affirm there is no charge of the same sign in the vicinity of the paramagnetic center. But the latter conclusion can be reliably drawn only when it is known with certainty that the probe charge is not neutralized by association with counterions whose concentration in solutions of biopolymers is often very high because of the use of buffer mixtures.

Table 4.32. Rate constants of spin exchange between ferricyanide anion and low-molecular nitroxide radicals [4.68]

Radical	$K_e \cdot 10^{-9}$ [l/mol · s]
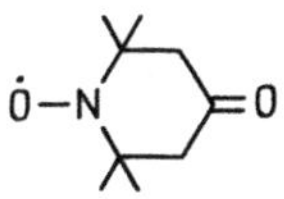	1.2
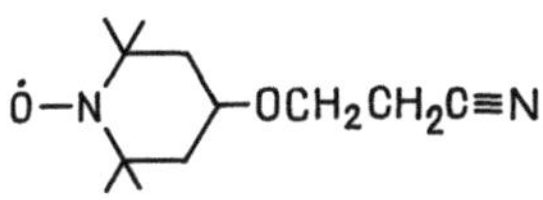	1.1
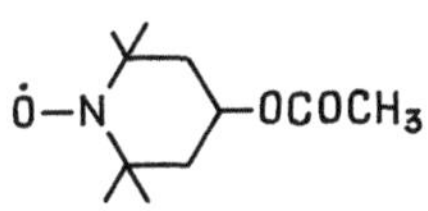	0.91
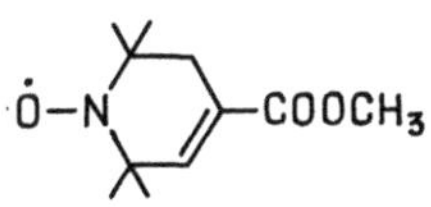	0.88
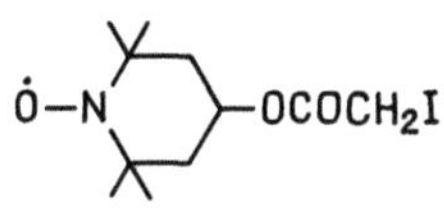	0.77
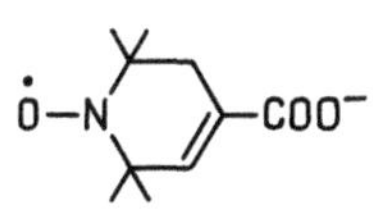	0.23

Compare, e.g., the values of K_e for the interaction of ferricyanide anion with spin-labelled proteins (see Table 4.31) and with low-molecular nitroxide radicals of various structure (Table 4.32). The values of K_e for the exchange with readily accessible labels of myoglobin, lysozyme, and albumin are about twice as small as those for the exchange with low-molecular radicals. Note that about the same difference could be expected due to the difference in the diffusion coefficient of the probe with respect to the radical groups of low- and high-molecular compounds. For low-molecular compounds this coefficient equals the sum $D_A + D_B$ of the diffusion coefficients of both particles involved in the exchange while for high-molecular ones it must be close to the diffusion coefficient D_A of the probe (i.e., about twice as small) since for a large polymer molecule the diffusion coefficient D_B is small ($D_B \ll D_A$).

Bearing this in mind, on the basis of comparison between the data listed in Tables 4.31 and 4.32 one can state that the molecules of myoglobin, lysozyme, and albumin do not practically create any steric hinderance for the collisions of the labels attached to them with the ferricyanide anion.

However, if the values of K_e for the interaction of the probes with the paramagnetic biopolymer centers are much smaller than for the interaction with their low-molecular analogs, then the interpretation of the reasons for this decrease should be approached with great care. Indeed, according to the results discussed in Sects.4.1-3 the small values of K_e may correspond to the weak exchange case, or may be associated with a small steric factor f and, in the case of charged probes, with the presence, near the paramagnetic centers, of electric charges of the same sign. The alteration of the rate of the probe diffusion in the vicinity of biopolymer molecule cannot be excluded either. Finally, without a careful investigation one cannot be sure that the broadening of ESR lines is due precisely to exchange, rather than to dipole-dipole interaction when the constants K_e are small. It seems that for biopolymer solutions, too, one can in many cases separate the contributions of different factors listed above by the same techniques as for more simple systems. To do this, it is necessary to study the dependence of the constants K_e for the exchange of a given biopolymer with a set of probes of different structures on the temperature, viscosity, and ionic strength of a solution. Unfortunately, this kind of research has not yet been carried out systematically. In principle, however, such research may be quite successful as was demonstrated above on the study of spin-labelled diffusion in membranes. By using the techniques described in Sect.3.3,4 the possibility of dipole-dipole broadening of ESR lines and of a weak exchange situation have been unambiguously excluded for these systems and thereby reliable data have been obtained on the kinetics of binary collisions in membranes.

4.4.3 "Development" of Individual ESR Spectra of Spin Labels via Spin Exchange

The experimentally observed ESR spectra of spin labels in biological systems are often a superposition of several signals from the labels of different nature or from one label in different environments. Paramagnetic probes introduced into such systems must cause exchange broadening of the ESR spectrum of the label, the lines from labels more accessible for collisions with probes being more broadened. This effect can be used to reveal ("develop") the lines of small intensity from the spin labels inaccessible for collisions on the background of the more intensive lines from the labels more accessible for collisions. Paramagnetic probes with a short electron relaxation time (e.g., ferricyanide anion, as well as nickel, cobalt, and rare earth cations) whose ESR spectrum is not observed are especially convenient for this purpose. The paramagnetic probe serves in this case as a broadening agent.

As an example illustrating this method, consider the ESR spectrum of TANONE added to dimyristoyllecithin (DML) phospholipid multilayers [4.78]. For a sample containing 20 water molecules per molecule of phospholipid the signal of free TANONE from the aqueous phase with rather narrow lines (Fig. 4.31) predominates in the ESR spectrum at 6°C. An addition of 0.5 M of Ni-Tris complex removes the signal from free TANONE which becomes strongly broadened due to the spin exchange with Ni. Thus, the spectrum retains only the wide lines from TANONE localized in the hydrocarbon phase. These lines are not broadened by exchange since no nickel penetrates into the hydrocarbon phase. The large width of the residual signal shows the mobility of the label in the hydrocarbon phase to be low. The study of the temperature dependence of the line width of this signal revealed a sharp change in mobility at about 25°C.

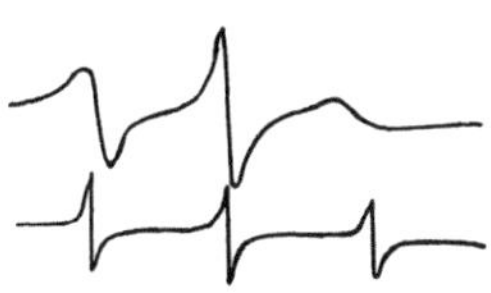

Fig. 4.31. Spectra of TANONE in the DML-20 water sample, at 6°C. The upper spectrum shows the residual signal of TANONE from hydrocarbon phase after the aqueous signal has been removed by the interaction with Ni-Tris. The lower spectrum shows that the aqueous line dominates in amplitude when no Ni-Tris is present

Another example is the study of the behavior of spin labels - nitroxide radicals - in various cells [4.79,80]. Nitroxide free radicals of the TANONE type can freely penetrate into the cell as they are soluble enough in the cell membranes separating the interior of the cell from the environment. To analyze the ESR spectrum of a spin label located within the cell, however, it is necessary to remove a far more intensive signal of the spin label located on the outside of the cell boundary. With this aim the salts $NiCl_2$ [4.79] and $K_3Fe(CN)_6$ [4.80] were used as broadening agents. Since Ni^{2+} or $Fe(CN)_6^{3-}$ did not penetrate the cell membrane (as shown by special experiments) the ESR spectrum of the nitroxide radical inside the cell was not changed. At the same time the signal of the radical in the aqueous environment was broadened due to spin exchange to such an extent that it became unobservable. The ESR signal thus isolated allowed the rotational diffusion mobility of the radical and hence the internal viscosity of the cell to be estimated from the line width. In the red blood cells [4.80] the internal viscosity turned out to be 4 times higher than the viscosity of water, while in some other cells (green alga, heier plant, mammalian) the ratio of viscosities exceeds an order of magnitude [4.79].

A vast body of information was also obtained from spin exchange study of the sarcoplasmic reticular vesicles from rabbit and lobster muscle [4.81]. The vesicles studied were about 1000 to 2000 Å in diameter and were bounded by a lipid-protein membrane. Three types of nitroxide spin labels were used: 1) labels located in the internal and external aqueous regions (e.g., TANONE); 2) labels located in the hydrocarbon of the membrane (spin-labelled analogs of fatty acids and cholesterol); and 3) spin labels covalently bound to sulfhydryl and amine groups of proteins which can be located in the membrane-water interface. The broadening agent in all the experiments was $NiCl_2$ which is water soluble and does not penetrate into the membrane.

When TANONE and other water-soluble radicals are used as spin labels, incorporation of $NiCl_2$ removes the ESR signal of the outside radicals just as was the case in the above experiments with cells. The remaining signal belongs to the radicals located inside the vesicle. Analysis of this signal shows the spin labels to rotate 30 to 50 times slower inside, than outside the vesicle. The spin exchange was also used to find out what is the spatial distribution of the spin labels inside the vesicle. For this purpose the dependence of the line width of the internal signal on the radical concentration was studied. The line broadening in vesicles turned out to be 3-6 times larger than that in a water-glycerol mixture of the same viscosity and with the same concentration of radicals. Hence it was concluded that the spin label is concentrated in some regions, possibly along the membrane inner surface.

The hydrophobic spin labels located in the membrane have a larger width of the ESR signal, i.e., they are strongly immobilized. Incorporation into the aqueous phase of $NiCl_2$ in sufficient concentrations (~ 1 M) removes about 60% of the signal intensity. The signal removed may be assumed [4.81] to belong to the labels located in the outer lipid layer of the membrane. This part of the labels appears to have a possibility of colliding with nickel ions of the outside water medium. As to the labels located in the inner layer, they do not interact with the broadening agent and their ESR spectrum is the same as that for the labels located in the outer layer in the absence of $NiCl_2$. From these experiments it was concluded that the mobility of the labels located in both the outer and the inner lipid layer is the same.

The labels covalently bound with the proteins are located on both the inner and the outer surface of the membrane. The broadening agent, $NiCl_2$, removes the narrower ESR signal of the outer labels and does not remove the broad signal of the strongly immobilized inner labels. This result agrees with the conclusion about the high viscosity of the medium inside the vesicle. Thus, the results considered have made it possible to obtain a detailed picture of the microviscosity distribution in vesicles [4.81].

The earlier investigations in which the spin exchange was used to reveal individual ESR spectra have been reviewed in [4.68].

4.4.4 Spin Exchange Titration

Exchange broadening of spin probe ESR spectra can be used for measuring the concentration of paramagnetic particles in biological systems. This method can be called spin exchange titration.

Kinetics of Reactions Involving Oxygen Molecules

The spin exchange titration allows one, e.g., to measure the kinetics of chemical and biochemical reactions accompanied by consumption or evolution of oxygen. Indeed, the spin exchange between paramagnetic oxygen and a spin probe transforms the ESR spectrum of the latter so that by analyzing the shape of this spectrum one can control the content of oxygen in the system. The practical possibilities of this method were demonstrated on an example of mitochondrial respiration [4.82]. A stable nitroxide radical

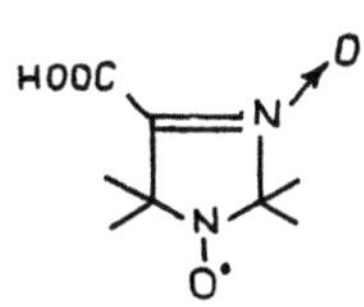

was used as a probe. The consumption of oxygen in the process of respiration was recorded in two ways.

In magnetically diluted solutions a resolved hfs of the ESR spectrum of the radical was observed which was due to the interaction of an unpaired electron with protons of methyl groups (Fig.4.32). The exchange broadening of the proton hfs components was determined by comparing the experimental ESR spectra with the computer simulated ones. Bearing in mind that exchange broadening is linear with oxygen concentration the broadening obtained can be used directly to determine the changes in the concentration of the dissolved oxygen.

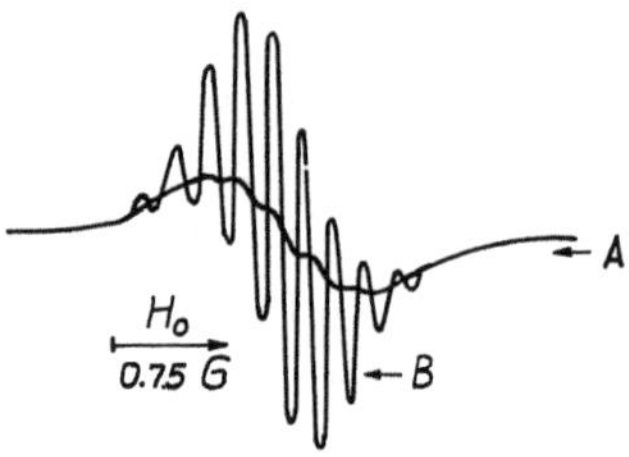

Fig. 4.32. Transformation of the nitroxide radical central ESR component due to the reduction of dissolved oxygen during mitochondrial respiration: (A) the central component before incubation with mitochondria, (B) the same component 5 min after incubation with mithochondrias, [R] = $7 \cdot 10^{-4}$ M [4.82]

In concentrated solutions when no proton hfs is observed, the kinetics of the oxygen reduction was obtained by recording the peak intensity of the nitrogen component of the ESR spectrum. When the line width is only slightly changed due to spin exchange the peak intensity displays in fact a linear dependence on the exchange broadening, i.e., on the concentration of oxygen.

Both techniques of recording yielded coincident results. The sensitivity of modern ESR spectrometers is probably sufficient to record in this way the respiration of one or several cells in a sample.

Coordination of Paramagnetic Ions by Macromolecules

The difference in the exchange broadening of the ESR spectra of probes by free and by complexed paramagnetic particles can be used to control the coordination of these particles by macromolecules [4.83]. Although spin exchange titration has so far been tried only on synthetic macromolecules it seems reasonable to describe it here, owing first, to its close intrinsic connection with the other technique described above and, secondly, to the possibility of applying it in studying biological macromolecules. The essence of the technique of spin exchange titration is demonstrated by Fig.4.33. Here the dependence of the ESR line width of the probe, 2,2,6,6-tetramethyl-4-oxypiperidine-

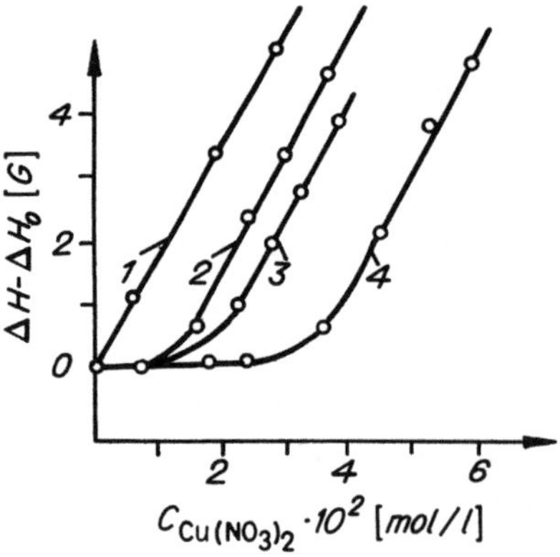

Fig. 4.33. The dependence of the ESR line width of R radical on the concentration of $Cu(NO_3)_2$ in the absence (1) and in the presence of poly-4-vinyl pyridine: C_{Py} = 0.14 (2), 0.21 (3), 0.36 mol/l (4)

1-oxyl free radical (R), on the concentration of copper(II) nitrate with and without additions of complexing polymer, poly-4-vinyl-pyridine is depicted. The titration of an aqueous solution of R with a solution of $Cu(NO_3)_2$ without polymer is seen to give a usual ESR line broadening of R linear in concentration due to spin exchange with an aquo complex of copper. However when a polymer is added to a solution of R, titration with a $Cu(NO_3)_2$ solution does not produce any broadening of the ESR lines of R till a definite concentration of $Cu(NO_3)_2$ is provided which depends on the concentration of added polymer. When this critical concentration is reached broadening linear in concentration recommences, with the same slope. Such a dependence of ΔH on C is due to the fact that the first portions of Cu(II) form with the polymer strong complexes $Cu(Py)_4$ (where Py is a pyridine fragment attached to the polymer) in which the copper atoms are screened by the polymer chain from collisions with R. From the ratio of concentrations of Cu(II) and of pyridine rings of the polymer in solution at the bending points of the curves $\Delta H = f(C)$ it is possible to determine this fraction of Py fragments of the polymer that can form $Cu(Py)_4$ complexes. It proved to be equal to $33 \pm 4\%$ of the total number of the pyridine rings in the polymer. Control measurements by an independent technique - spectrophotometric titration - yielded the same result.

A specific feature of the spin exchange titration technique is its selective sensitivity to the complexes that are buried inside the polymer globule and hence are inaccessible to the probe molecules. This offers a possibility of determining the concentrations of bulk and surface paramagnetic centers when combining this technique with other methods of quantitative analysis. For example, in the case of poly-4-vinylpyridine the coincidence of the data obtained by spin exchange and spectrophotometric titration shows that the fraction of copper atoms forming the surface complexes cannot exceed the relative error of measurement, i.e., 1/8.

4.4.5 Other Applications

Spin exchange as a subsidiary technique permitting one to broaden the lines of paramagnetic labels was used [4.84] in a new highly sensitive method of studying the immune lysis of liposomes and erythrocyte ghosts. The ghosts or liposomes were loaded with a concentrated solution of a water-soluble, membrane-impermeable spin label. Owing to exchange interaction the ESR signal of the spin label was strongly broadened. As a result of the immune lysis caused by the interaction between the antigen (egg albumin) associated with the membrane surface and the antibody in the presence of complement the radicals were released into a dilute solution which was recorded by a sharp increase in the amplitude of the ESR signal. The method has a high sensitivity: the lysis of a concentration of human erythrocyte ghosts equivalent to 10^{-15} M (6×10^8 ghosts per liter) is detected with the signal-to-noise ratio of no less than 10/1 in a volume of about 20 μl.

Spin exchange can be successfully used to investigate the penetration of paramagnetic substances through the membrane. It was used, e.g., to study the diffusion through the membrane of the $Cu(Lis)_2$ complex [4.85] which is shown to be an inhibitor of reactions catalyzed by cytochrome P-450 located inside the membrane. In this investigation liposomes of egg cytochrome were loaded with the water-soluble radical Fremy salt. These radicals took no part in the exchange interaction with $Cu(Lis)_2$ dissolved in outside water. This result precludes the possibility of free diffusion of the $Cu(Lis)_2$ complex through the membrane. The complex appears to operate as an inhibitor by incorporating itself into the outer layer of the membrane.

Thus, spin exchange studies yield important information on the rates of collisions of molecules and their fragments in biological systems. Here, just as in the case of simpler systems the rate constants determined by this method automatically reflect the influence on the collisions rate of numerous factors whose role often cannot be even roughly estimated a priori, such as the ordering of the medium in which molecules diffuse, its inhomogeneity, solvation effects, ionic association, etc.

4.5 Conclusion

At present spin exchange is most widely used to measure the number of binary collisions in solutions. And what is of particular importance is that such measurements can be made in fairly complex systems for which even rough theoretical estimates are in many cases so far impossible. These systems include the solutions of particles whose size is smaller than that of solvent molecules, multicomponent solutions, solutions of polymers and polymeric matrices, and biological systems.

Spin exchange has also proved sensitive to such details of collisions as, e.g., steric restrictions. This provides an opportunity for using spin exchange to study this factor which is important for chemical reactions. So far the investigations in this direction are not numerous but their results are very promising. Appreciable steric restrictions manifest themselves, e.g., in studying spin exchange between stable free radicals and simple complexes of transition metals. Particularly essential steric restrictions, however, are observed when radicals collide with active centers of biopolymers, such as hemoglobin or cytochrome C.

Spin exchange offers important possibilities in the study of electrostatic interactions of ions during their collisions. The electrostatic attraction or repulsion is known to essentially affect the rate of ionic reactions. Experiments on spin exchange make it possible to estimate quantitatively the role of electrostatic effects, even under the conditions where their theoretical estimation is practically infeasible. These experiments allow thereby to take into account the dependence of the collision rate of charged particles on such factors as asymmetric distribution of charges, association with counterions, screening by ionic atmosphere, discrete molecular structure of solvent, etc.

Spin exchange studies have resulted in the development of a theory which was experimentally verified. This theory describes quantitatively the dependence of the rate of spin state changes during collision on such characteristics as energy of exchange interaction, values of electron spins of interacting particles, and rate of electron spin relaxation. After some modifications this theory can also be extended to other elementary processes involving spin state changes in particles. The first example of this kind is the ortho-para-conversion of positronium. This theory may be spread to other important processes occurring during collisions of paramagnetic particles, e.g., to the exchange quenching of luminescence.

Of the chemical reactions those that are the closest to spin exchange in their nature are outersphere electron transfer reactions. Experiments on spin exchange allow the quantitative estimation of the dependence of the reaction rate of outer sphere electron transfer on such important factors as electrostatic interaction between reagents at the moment of collision and steric restrictions. In paramagnetic particles the orbitals of unpaired electrons are simultaneously both the lower vacant and the upper occupied ones. Outer sphere electron transfer reactions, therefore require an overlap of the same orbitals as does spin exchange. Under these conditions the steric factors for the two processes must indeed be close to each other.

The above coincidence of orbitals taking part in electron transfer and spin exchange allows the use of spin exchange to find out which of the two basically possible mechanisms - adiabatic or nonadiabatic - is realized in a particular outer sphere electron transfer reaction. The solution of this problem is of primary significance for the theory of these processes. Therefore it should be stated here in more detail.

Electron transfer, like spin exchange, is provided by the overlap of orbitals of colliding reagents. However, the matrix elements of transition for the two processes are different. Spin exchange between two paramagnetic particles is determined by the exchange integral

$$J = J_0 S_{ab}^2$$

while electron transfer by the resonance integral

$$V = V_0 S_{ab} \ .$$

In these expressions S_{ab} is the overlap integral; J_0 and V_0 are the numerical coefficients which, in contrast to S, are nearly independent of the nature of reagents and can be estimated on the basis of quantum chemical calculations.

The efficiency of spin exchange is determined by an action performed by the exchange integral within the contact time of the partners. The efficiency of an electron transfer reaction is determined by an action which the resonance integral performs within a short time t during which the medium polarization and the atom configuration will create a favorable condition for electron transfer. An electron transfer reaction proceeds adiabatically if $Vt > 1$. If the opposite condition, $Vt < 1$, is fulfilled, the reaction is nonadiabatic. The above relations for J and V enable one to estimate the values of the resonance integral V for an electron transfer reaction using for this the value of the exchange integral obtained from experiments on spin exchange.

For some processes in liquids spin exchange is not only an instrument of research, but also their essential component. To these belong chemically induced dynamic electron polarization (CIDEP) and chemically induced dynamic nuclear polarization (CIDNP). In the case of CIDEP it is spin exchange that provides the creation of nonequilibrium population of spin sublevels, so that in the absence of spin exchange no electron polarization is possible. As for the CIDNP effect, here, too, spin exchange can also play an essential role though it is not the necessary condition of the effect.

Thus, the studies on the mechanism of spin exchange have made it possible to create a new sensitive method for investigating the collision of molecules in liquids. This method has yielded useful information for understanding the nature of a number of important chemical and physical phenomena.

List of Frequently Used Notations and Abbreviations

Kinetics of Spin Exchange

K_e	Rate constant of spin exchange
K_D	Rate constant of diffusion collisions in solution
K_b	Rate constant of concentrational broadening of ESR lines
D_A, D_B	Diffusion coefficients of particles A, B.
$D_{AB} = D_A + D_B$	
η	Viscosity of solution
R_A, R_B	Van der Waals radii of particles A, B.
$R_{AB} = R_A + R_B$	
C	Concentration of paramagnetic particles
p	Efficiency of spin exchange during a collision
f	Steric factor for spin exchange

Quantum Mechanics

J	Exchange integral
S_{ab}	Overlap integral of orbitals a and b
$\hat{H}$	Spin Hamiltonian
ρ_a, ρ_b	Spin densities on atoms a, b
S	Electron spin of paramagnetic particle
I	Nuclear spin

ESR Spectroscopy

ΔH	ESR line width between the points of maximum slope [Gauss]
ΔH_0	ESR line width between the points of maximum slope in the absence of spin exchange [Gauss]
ΔH_e	Spin exchange contribution to the ESR line width between the points of the maximum slope [Gauss]
$\Delta H_{1/2}$	Half width of ESR line at half height [Gauss]
ΔH_{hf}	Hyperfine interaction constant [Gauss]

a	Hyperfine interaction constant [rad/s]
T_1	Longitudinal relaxation time of electron spin
T_2	Transverse relaxation time of electron spin
g	g Value of paramagnetic particle
ESR	Electron spin resonance
hfi	Hyperfine interaction

Chemical Formulae

TANONE	2,2,6,6-tetramethyl-4-oxopiperidine-1-oxyl	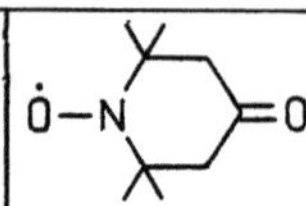
DTBN	di-tert-butyl-nitroxyl	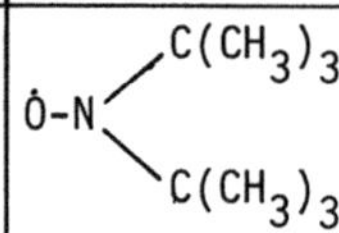
DPPH	α, α'-diphenyl-β-picryl-hydrazyl	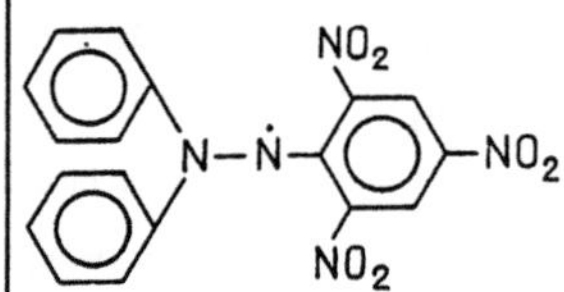
BDC	bisdiphenylchromium cation	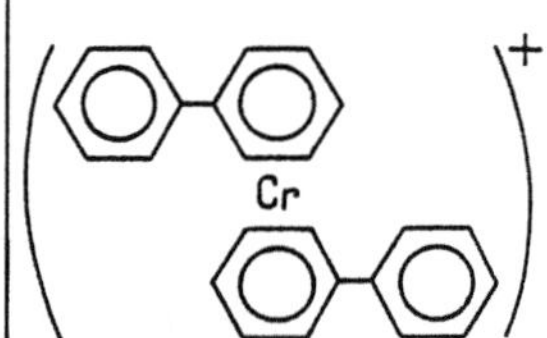
en	ethylene diamine	$NH_2CH_2CH_2NH_2$
acac	acethylacetonate anion	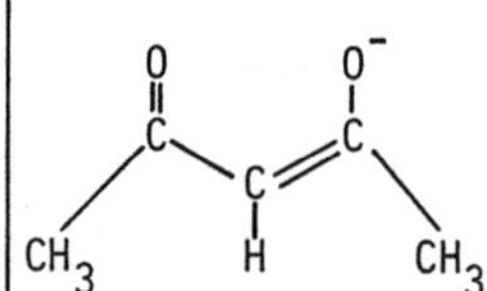

References

Chapter 1

1.1 L.D. Landau, Ye.M. Lifschitz: *Quantum Mechanics* (Fizmatgiz, Moscow 1963) Chap. 1, pp. 58-61, Chap. 9, pp. 255-260
1.2 R. McWeeny, B.T. Sutcliffe: *Methods of Molecular Quantum Mechanics* (Academic Press, New York 1969) Chaps. 3,4
1.3 A. Carrington, A.D. McLachlan: *Introduction to Magnetic Resonance* (Harper and Row, New York 1967)
1.4 J.E. Wertz, J.R. Bolton: *Electron spin resonance* (McGraw-Hill, New York 1972)
1.5 E.F. Caldin: *Fast Reactions in Solution* (Blackwell Scientific Publications, Oxford 1964)
1.6 E.A. Moelwyn-Hughes: *The Chemical Statics and Kinetics of Solutions* (Academic Press, New York 1971)
1.7 Ye.K. Zavoisky: Doctor of Sciences Thesis, Moscow FIAN (1944); J. Phys. USSR, *9*, 245 (1945)
1.8 C.J. Gorter, J.H. Van Vleck: Phys. Rev. *72*, 1128 (1947)
1.9 J.H. Van Vleck: Phys. Rev. *74*, 1168 (1948)
1.10 D.M.S. Bagguley, J.H.E. Griffiths: Nature *162*, 538 (1948)
1.11 M.H.L. Pryce: Nature *162*, 539 (1948)
1.12 M.A. Garstens, S.H. Liebson: J. Chem. Phys. *20*, 1647 (1952)
1.13 P.W. Anderson, P.R. Weiss: Rev. Mod. Phys. *25*, 269 (1953)
1.14 P.W. Anderson: J. Phys. Soc. Jpn. *9*, 315 (1954)
1.15 J.P. Wittke, R.H. Dicke: Phys. Rev. *103*, 620 (1956)
1.16 G.E. Pake, T.R. Tuttle, Jr.: Phys. Rev. Lett. *3*, 423 (1959)
1.17 D. Kivelson: J. Chem. Phys. *33*, 1094 (1960)
1.18 K.M. Salikhov, A.B. Doktorov, Yu.N. Molin, K.I. Zamaraev: J. Magn. Reson. *5*, 189 (1971)
1.19 E.G. Rozantsev: *Free Iminoxyl Radicals* (Khimiya, Moscow 1970)
1.20 A.L. Buchachenko, A.M. Wasserman: *Stable Radicals* (Khimiya, Moscow 1973)
1.21 G.I. Likhtenshtein: *The Method of Spin Labels in Molecular Biology* (Nauka, Moscow 1974)

Chapter 2

2.1 S.A. Al'tshuler, B.M. Kozyrev: *Electron Paramagnetic Resonance of Compounds of Intermediate Group Elements* (Nauka, Moscow 1972) Chaps. 3,4, pp. 54-221
2.2 S.V. Vonsovsky, B.V. Karpenko: "Application of the Bogolubov-Tyablikov Method to Problems of Exchange Interaction in Nonmetallic Magnetically Ordered Crystals", in *Problems of Theoretical Physics* (Nauka, Moscow 1969) pp. 294-311
2.3 P.D. Dacre, R. McWeeny: Proc. Roy. Soc. *A317*, 435 (1970)
2.4 G.A. Smolensky, V.V. Lemanov, G.M. Nedlin, M.P. Petrov, R.V. Pisarev:

Physics of Magnetic Dielectrics (Nauka, Leningrad 1974) Chap. 1
2.5 G.M. Zhidomirov, P.V. Schastnev, N.D. Chuvylkin: *Quantum-Chemical Calculations of Magnetic Resonance Parameters* (Nauka, Novosibirsk 1978) Chap. 5, pp. 308-335
2.6 L.D. Landau, Ye.M. Lifshits: *Quantum Mechanics* (Fizmatgiz, Moscow 1963) Chap. 1, pp. 58-61, Chap. 9, pp. 255-260
2.7 R. McWeeny, F. Yonezawa: J. Chem. Phys. Suppl. *43*, 120 (1965)
2.8 L.P. Gor'kov, L.P. Pitayevsky: Dokl. Akad. Nauk SSSR *151*, 822 (1963)
2.9 B.M. Smirnov: *Asymptotic Methods in the Theory of Atomic Collisions* (Atomizdat, Moscow 1973) pp. 1-294
2.10 Per-Olov Lödwin: Rev. Mod. Phys. *34*, 80 (1962)
2.11 R. McWeeny, B.T. Sutcliffe: *Methods of Molecular Quantum Mechanics* (Academic Press, New York 1969) Chaps. 3,4
2.12 P.V. Schastnev, K.M. Salikhov: Teor. Eksp. Khim. *9*, 291 (1973)
2.13 R.N. Musin, P.V. Schastnev: Zh. Strukt. Khim. *17*, 411 (1976)
2.14 P.W. Anderson: Phys. Rev. *79*, 350 (1950)
2.15 R.N. Musin, P.V. Schastnev: Chem. Phys. Lett. *51*, 187 (1977)
2.16 Nai Li Huang, R. Orbach: Phys. Rev. *154*, 487 (1967)
2.17 A.V. Zvarykina, V.B. Strykov, S.Ya. Umansky, D.N. Fedutin, R.P. Shibayev, E.G. Rozantsev: Dokl. Akad. Nauk SSSR *216*, 1091 (1974)
2.18 R.N. Musin, P.V. Schastnev: Zh. Strukt. Khim. *17*, 419 (1976)
2.19 J. Owen: J. Appl. Phys. Suppl. *33*, 355 (1962)
2.20 J.O. Hirschfelder, Ch.F. Curtiss, R.B. Bird: *Molecular Theory of Gases and Liquids* (John Wiley, New York; Chapman and Hall, Ltd., London 1954) Chap. 14
2.21 Yu.N. Molin: "Distribution of Unpaired Electrons in Paramagnetic Complexes and Elementary Processes Involving Spin State Changes", Doctor of Sciences Thesis, Novosibirsk (1970)
2.22 A. Carrington, A.D. McLachlan: *Introduction to Magnetic Resonance with Applications to Chemistry and Chemical Physics* (Harper and Row, New York 1967) Chap. 6
2.23 K.M. Salikhov: The Kinetics of Processes Induced by Spin-Spin Interactions of Particles in Magnetically Diluted Systems", Doctor of Sciences Thesis, Kazan (1974)
2.24 H.A. Kramers: Physica *1*, 182 (1934)
2.25 N.L. Huang: Phys. Rev. *164*, 636 (1967)
2.26 A.L. Buchachenko, R.Z. Sagdeev, K.M. Salikhov: *Magnetic and Spin Effects in Chemical Reactions* (Nauka, Novosibirsk 1978) Chaps. 1,2, pp. 7-54
2.27 G.E. Pake, T.R. Tuttle, Jr.: Phys. Rev. Lett. *3*, 423 (1959)
2.28 N.M. Bazhin, K.M. Salikhov: *ESR. Relaxation of Free Radicals in Liquids* (Novosibirsk Univ., Novosibirsk 1974) pp. 1-128
2.29 P.W. Anderson, P.R. Weiss: Rev. Mod. Phys. *25*, 269 (1953)
2.30 P.W. Anderson: J. Phys. Soc. Jpn. *9*, 315 (1954)
2.31 C.P. Slichter: *Principles of Magnetic Resonance*, Springer Series in Solid State Sciences, Vol. 1 (Springer, Berlin, Heidelberg, New York 1978)
2.32 Ch.S. Johnson, Jr.: "Chemical Rate Processes and Magnetic Resonance", in *Advances in Magnetic Resonance*, Vol. 1 (Academic Press, New York 1965) pp. 33-102
2.33 A.I. Burshtein: *Quantum Kinetics, A Course of Lectures* (Novosibirsk Univ., Novosibirsk 1968) pp. 1-230
2.34 K.M. Salikhov, A.G. Semenov, Yu.D. Tsvetkov: *Electron Spin Echo and Its Application* (Nauka, Novisibirsk 1976) Chap. 4, pp. 195-236
2.35 H.M. McConnell: J. Chem. Phys. *28*, 430 (1958)
2.36 M. Loeve: *Probability Theory* (D. Van Nostrand Company Inc., Princeton, New Jersey 1955) Chap. 12
2.37 A.I. Burshtein, Yu.S. Oseledchik: Zh. Eksp. Teor. Fiz. *51*, 1071 (1966)
2.38 I.W. Emsley, J. Feeny, L.H. Sutcliffe: *High Resolution Nuclear Magnetic*

Resonance Spectroscopy, Vol. 1 (Pergamon Press, London 1965) Chap. 9, pp. 481-502
2.39 V.L. Smirnov: *Course of Higher Mathematics*, Vol. 3, Part 1 (Fizmatgiz, Moscow 1958)
2.40 A.I. Burshtein: Zh. Eksp. Teor. Fiz. *54*, 1120 (1968)
2.41 E.L. Wolf: Phys. Rev. *142*, 555 (1966)
2.42 K.M. Salikhov, A.B. Doktorov: "Theory of Steady-State Saturation of ESR Spectra with Spectral Diffusion Taken into Account", in *Paramagnetic Resonance 1944-1969*, Part 2 (Kazan 1971) pp. 122-126
2.43 J.H. Freed: J. Chem. Phys. *43*, 2312 (1965)
2.44 M.P. Eastman, R.G. Kooser, M.R. Das, J.H. Freed: J. Chem. Phys. *51*, 2690 (1969)
2.45 J.H. Freed: J. Phys. Chem. *71*, 38 (1967)
2.46 A.M. Portis: Phys. Rev. *91*, 1071 (1953)
2.47 T.G. Castner, Jr.: Phys. Rev. *115*, 1506 (1959)
2.48 O.P. Zhidkov, Ya.S. Lebedev, A.I. Mikhailov, B.N. Provotorov: Teor. Eksp. Khim. *3*, 240 (1967)
2.49 J.S. Hyde: "Electron-Electron Double Resonance", in *Free-Radical States in Chemistry*, ed. by L.A. Blumenfeld, Yu.N. Molin (Nauka, Novosibirsk 1972) pp. 24-38
2.50 V.A. Bendersky, L.A. Blumenfeld, P.A. Stunzhas, Ye.A. Sokolov: "Double Electron-Electron Resonance. Interaction of Triplet Excitons in Ion-Radical Salts", in *Free-Radical States in Chemistry* (Nauka, Novosibirsk 1972) pp. 39-47
2.51 M. Huisjen, J.S. Hyde: J. Chem. Phys. *6*, 1682 (1974)
2.52 Ya.S. Lebedev, V.I. Muromtsev: *ESR and Relaxation of Stabilized Radicals* (Khimiya, Moscow 1972) Chaps. 4,5 pp. 109-194
2.53 V.A. Atsarkin, M.I. Rodak: Usp. Fiz. Nauk *107*, 3 (1972)
2.54 M. Nechtschein, J.S. Hyde: Phys. Rev. Lett. *24*, 672 (1970)
2.55 J.S. Hyde, J.C.W. Chien, J.H. Freed: J. Chem. Phys. *48*, 4211 (1968)
2.56 P.A. Stunzhas, V.A. Bendersky: Opt. Spektrosk. *30*, 1041 (1971)
2.57 A.D. Milov, K.M. Salikhov, Yu.D. Tsvetkov: Chem. Phys. Lett. *8*, 523 (1971)
2.58 R.Z. Sagdeev: "Far Hyperfine Interactions in Saturated Systems", Candidate of Sciences Thesis, Kazan (1969)
2.59 H.C. Berg: Phys. Rev. *A137*, 1621 (1965)
2.60 Ye.B. Gordon, A.N. Ponomarev, V.L. Tal'roze: Khim. Vysok. Energ. *2*, 512 (1968)
2.61 D. Kivelson: J. Chem. Phys. *33*, 1094 (1960)
2.62 L.I. Schiff: *Quantum Mechanics* (McGraw-Hill, New York 1955) Chap. 8, Sect. 31
2.63 R.C. Johnson, R.E. Merrifield: Phys. Rev. *B1*, 896 (1970)
2.64 P.W. Atkins, G.T. Evans: Mol. Phys. *29*, 921 (1975)
2.65 R. Kaptein, J.L. Oosterhoff: Chem. Phys. Lett. *4*, 195 (1969)
2.66 T.R. Waite: J. Chem. Phys. *28*, 103 (1958)
2.67 K.M. Salikhov: Teor. Eksp. Khim. *10*, 310 (1974)
2.68 J.D. Currin: Phys. Rev. *126*, 1995 (1962)
2.69 R.M. Lynden-Bell: Mol. Phys. *8*, 71 (1964)
2.70 J.H. Freed: J. Chem. Phys. *45*, 3452 (1966)
2.71 C.S. Johnson: Mol. Phys. *12*, 25 (1967)
2.72 K.M. Salikhov, A.T. Nikitayev, G.A. Senyukova, K.I. Zamaraev: Teor. Eksp. Khim. *7*, 619 (1971)
2.73 K.M. Salikhov, A.B. Doktorov: "Exchange Broadening of the ESR Spectra Lines in Diluted Solutions", in *Magnetic Resonance and Related Phenomena*, Proc. XVI Congr. AMPERE, Bucharest 1970, pp. 399-403
2.74 V.Yu. Zitserman: Mol. Phys. *20*, 1005 (1971)
2.75 A.L. Buchachenko, A.M. Wasserman: *Stable Radicals* (Khimiya, Moscow 1973)

2.76 M.L. Berkovich, K.M. Salikhov: Teor. Eksp. Khim. *9*, 586 (1973)
2.77 G.I. Skubnevskaya, K.M. Salikhov, L.M. Smirnova, Yu.N. Molin: Kinet. Katal. *11*, 888 (1970)
2.78 K.M. Salikhov, A.B. Doktorov, Yu.N. Molin, K.I. Zamaraev: J. Magn. Reson. *5*, 189 (1971)
2.79 L.C. Balling, R.J. Hanson, F.M. Pipkin: Phys. Rev. *A133*, 607 (1964)
2.80 L.C. Balling, F.M. Pipkin: Phys. Rev. *A136*, 46 (1964)
2.81 N. Edelstein, A. Kwok, A.H. Maki: J. Chem. Phys. *41*, 3473 (1964)
2.82 N.N. Bogolubov: *Problems of the Dynamic Theory in Statistical Physics* (Fizmatgiz, Moscow, Leningrad 1946) pp. 1-200
2.83 K.M. Salikhov: Teor. Eksp. Khim. *15*, 483 (1979)
2.84 S. Chandrasekhar: Rev. Mod. Phys. *15*, 1 (1943)
2.85 K.M. Salikhov: Teor. Eksp. Khim. *13*, 732 (1977)
2.86 J.B. Pedersen: Chem. Phys. Lett. *52*, 333 (1977)
2.87 R.M. Noyes: J. Am. Chem. Soc. *78*, 5486 (1956)
2.88 A.B. Doktorov: Candidate of Sciences Thesis, Novosibirsk (1973)
2.89 V.N. Parmon, G.M. Zhidomirov: Mol. Phys. *27*, 367 (1974)
2.90 G.R. Luckhurst: Mol. Phys. *10*, 543 (1966)
2.91 V.N. Parmon: Candidate of Sciences Thesis, Moscow (1975)
2.92 G.M. Zhidomirov, Ya.S. Lebedev, S.N. Dobryakov, N.Ya. Shteinshneider, A.K. Chirkov, V.A. Gubanov: *Interpretation of Complex ESR Spectra* (Nauka, Moscow 1975)
2.93 Ch.S. Johnson, Jr.: J. Chem. Phys. *41*, 3277 (1964)
2.94 N.N. Korst, A.V. Lazarev: Physica *42*, 31 (1969)
2.95 G.I. Likhtenstein: *Methods of Spin Labels in Molecular Biology* (Nauka, Moscow 1974) pp. 1-255
2.96 M.V. Artamonova, Ch.M. Briskina, A.I. Burshtein, L.D. Zusman, A.G. Skleznev: Zh. Eksp. Teor. Fiz. *62*, 863 (1972)
2.97 A.B. Doktorov: Physica *90A*, 109 (1978)
2.98 K.I. Zamaraev, Yu.N. Molin, K.M. Salikhov: *Spin Exchange. Theory and Physicochemical Application* (Nauka, Novosibirsk 1977) Chap. 1, pp. 113, 114

Chapter 3

3.1 J.G. Powles, M.H. Mosley: Proc. Phys. Soc. *78*, 370 (1961)
3.2 A.T. Nikitayev: Candidate of Sciences Thesis, Moscow (1972)
3.3 V.N. Parmon, A.T. Nikitayev, G.M. Zhidomirov, K.I. Zamaraev: Zh. Strukt. Khim. *13*, 607 (1972)
3.4 Ya.C. Lebedev, V.V. Voyevodsky (eds.): *Atlas of ESR Spectra* (Nauka, Moscow 1962)
3.5 J.C. Lang, J.H. Freed: J. Chem. Phys. *56*, 4103 (1972)
3.6 Ya.S. Lebedev, V.I. Muromtsev: *ESR and Relaxation of Stabilized Radicals* (Khimiya, Moscow 1972)
3.7 K.I. Zamaraev, A.T. Nikitaev: "Study of the Electron Spin Exchange Mechanism in Liquid and Solid Solutions", in *Free-Radical States in Chemistry* (Nauka, Novosibirsk 1972) pp. 102-111
3.8 T.A. Miller, R.N. Adams, P.M. Richards: J. Chem. Phys. *44*, 4022 (1966)
3.9 P.W. Anderson, P.R. Weiss: Rev. Mod. Phys. *25*, 269 (1953)
3.10 D. Kivelson: J. Chem. Phys. *33*, 1094 (1960)
3.11 R.Kh. Timerov, K.A. Valiev: Zh. Eksp. Teor. Fiz. *41*, 1566 (1962)
3.12 K.I. Zamaraev, V.V. Voyevodsky: Dokl. Akad. Nauk SSSR *169*, 385 (1966)
3.13 K.I. Zamaraev, A.T. Nikitaev: Dokl. Akad. Nauk SSSR *188*, 1084 (1969)
3.14 A.Ya. Sychev, K.I. Zamaraev: Zh. Strukt. Khim. *11*, 538 (1970)
3.15 K.I. Zamaraev: Doctor of Sciences Thesis, Moscow (1972)
3.16 M.T. Jones: J. Chem. Phys. *38*, 2892 (1963)
3.17 G.I. Skubnevskaya: Candidate of Sciences Thesis, Novosibirsk (1972)
3.18 V.N. Parmon, A.T. Nikitaev, G.M. Zhidomirov, K.I. Zamaraev: Zh. Strukt.

Khim. *13*, 400 (1972)
3.19 A.M. Portis: Phys. Rev. *91*, 1071 (1953)
3.20 A.D. Milov, K.M. Salikhov, Yu.D. Tsvetkov: Chem. Phys. Lett. *8*, 523 (1971)
3.21 M.P. Eastman, R.G. Kooser, M.R. Das, J.H. Freed: J. Chem. Phys. *51*, 2690 (1969)
3.22 R.D. Rataiczak, M.T. Jones: J. Chem. Phys. *56*, 3898 (1972)
3.23 J.S. Hyde, T. Sarna: J. Chem. Phys. *68*, 4439 (1978)
3.24 K.M. Salikhov, A.G. Semenov, Yu.D. Tsvetkov: *Electron Spin Echo and Its Application* (Nauka, Novosibirsk 1976) Chap. 4, pp. 195-235
3.25 L.L. Makarshin, V.M. Berdnikov: Teor. Eksp. Khim. *12*, 494 (1976)
3.26 J.S. Hyde, J.C.W. Chien, J.H. Freed: J. Chem. Phys. *48*, 4211 (1968)
3.27 J.S. Hyde: "Electron-Electron Double Resonance", in *Free-Radical States in Chemistry*, ed. by L.A. Blumenfeld, Yu.N. Molin (Nauka, Novosibirsk 1972) pp. 24-38
3.28 A. Abragam: *The Principles of Nuclear Magnetism* (Clarendon Press, Oxford 1961)
3.29 H. Pfeifer: Ann. Phys. *8*, 1 (1961)
3.30 L.P. Hwang, J.H. Freed: J. Chem. Phys. *63*, 4017 (1975)
3.31 J.H. Freed: J. Chem. Phys. *68*, 4034 (1978)
3.32 K.I. Zamaraev, A.T. Nikitaev, G.A. Senyukova: Kinet. Katal. *13*, 54 (1972)
3.33 O.A. Ainisimov, A.T. Nikitaev, K.I. Zamaraev, Yu.N. Molin: Teor. Eksp. Khim. *7*, 682 (1971)
3.34 K.M. Salikhov, A.B. Doktorov, Yu.N. Molin, K.I. Zamaraev: J. Magn. Reson. *5*, 189 (1971)

Chapter 4

4.1 A.M. North: *The Collision Theory of Chemical Reaction in Liquids* (Methen Co., Ltd., New York 1964) pp. 1-145
4.2 E.A. Moelwyn-Hughes: *Physical Chemistry* (Pergamon Press, London 1961)
4.3 S.G. Entelis, R.P. Tiger: *The Kinetics of Reactions in Liquid State* (Khimiya, Moscow 1973)
4.4 E.F. Caldin: *Fast Reactions in Solution* (Blackwell Scientific Publications, Oxford 1964)
4.5 V.I. Goldansky: *The Physical Chemistry of the Positron and Positronium* (Nauka, Moscow 1968)
4.6 K.M. Salikhov, A.T. Nikitaev, G.A. Senyukova, K.I. Zamaraev: Teor. Eksp. Khim. *7*, 619 (1971)
4.7 N. Edelstein, A. Kwok, A.N. Maki: J. Chem. Phys. *41*, 3473 (1964)
4.8 W. Plachy, D. Kivelson: J. Chem. Phys. *47*, 3312 (1967)
4.9 M.P. Eastman, R.G. Kooser, M.R. Das, J.H. Freed: J. Chem. Phys. *51*, 2690 (1969)
4.10 M.P. Eastman, G.V. Bruno, J.H. Freed: J. Chem. Phys. *52*, 2511 (1970)
4.11 L.A. Buchachenko, L.S. Troitskaya: Izv. Akad. Nauk SSSR, Seria Khim. *4*, 602 (1966)
4.12 A.L. Kovarskii, A.M. Wasserman, A.L. Buchachenko: J. Magn. Reson. *7*, 225 (1972)
4.13 A.I. Kokorin, K.I. Zamaraev: Zh. Fiz. Khim. *46*, 1853 (1972)
4.14 O.A. Anisimov, A.T. Nikitaev, K.I. Zamaraev, Yu.N. Molin: Teor. Eksp. Khim. *7*, 682 (1971)
4.15 T.A. Miller, R.N. Adams, P.M. Richards: J. Chem. Phys. *44*, 4022 (1966)
4.16 J.C. Lang, J.H. Freed: J. Chem. Phys. *56*, 4103 (1972)
4.17 Yu.B. Grebenshchikov, G.I. Likhtenstein, V.P. Ivanov, E.G. Rozantsev: Molek. Biol. *6*, 498 (1972)
4.18 S. Ablett, M.D. Barratt, F. Franks: J. Sol. Chem. *4*, 797 (1975)
4.19 A.L. Buchachenko, A.M. Wasserman: *Stable Radicals* (Khimiya, Moscow 1973)
4.20 A.D. Milov, A.B. Melnik, Yu.D. Tsvetkov: Teor. Eksp. Khim. *11*, 790 (1975)
4.21 G.E. Pake, T.R. Tuttle: Phys. Rev. Lett. *3*, 423 (1959)

4.22 G.I. Skubnevskaya, Ye.Ye. Zayev, R.I. Zusman, Yu.N. Molin: Dokl. Akad. Nauk SSSR *170*, 386 (1966)
4.23 G.I. Skubnevskaya, Yu.N. Molin: Kinet. Katal. *13*, 1383 (1972)
4.24 G.I. Skubnevskaya: Candidate of Sciences Thesis, Novosibirsk (1972)
4.25 J. Owen, J.H.M. Thornley: Rep. Prog. Phys. *29*, 675 (1966)
4.26 Yu.N. Molin: Doctor of Sciences Thesis, Novosibirsk (1970)
4.27 G.I. Skubnevskaya, Yu.N. Molin: Kinet. Katal. *8*, 1192 (1967)
4.28 L.L. Makarshin, V.M. Berdnikov: Teor. Eksp. Khim. (in press)
4.29 K.I. Zamaraev, A.I. Kokorin: Zh. Fiz. Khim. *11*, 2855 (1976)
4.30 Yu.B. Grebenshchikov: Candidate of Sciences Thesis, Moscow (1973)
4.31 L.L. Makarshin, V.M. Berdnikov: Teor. Eksp. Khim. *12*, 494 (1976)
4.32 G.A. Senyukova: Graduation Thesis, Mosk. Fiz.-Tekhn. Inst. - Inst. Khim. Fiz. Akad. Nauk SSSR, Moscow (1969)
4.33 Yu.I. Naberukhin, V.A. Rogov: Usp. Khim. *40*, 369 (1971)
4.34 A.Yu. Shaulov, A.M. Wasserman, A.M. Brodsky, N.S. Yenikolopyan: Dokl. Akad. Nauk SSSR *196*, 888 (1971)
4.35 A.Yu. Shaulov, N.I. Andreyeva, A.G. Sklarova, A.L. Buchachenko, N.S. Yenikolopyan, Yu.Kh. Shaulov: Zh. Eksp. Teor. Khim. *63*, 157 (1972)
4.36 R.M. Barrer: J. Phys. Chem. *61*, 178 (1957)
4.37 A.F. Guk, K.I. Zamaraev, Z.G. Kozlova, Ya.S. Lebedev, A.T. Nikitaev, V.F. Tsepalov: Kinet. Katal. *13*, 1164 (1972)
4.38 A.M. Wasserman, G.V. Korolev, A.L. Kovarsky, A.I. Malakhov, B.R. Smirnov, T.V. Sosnina, A.M. Fokin: Izv. Akad. Nauk SSSR, Ser. Khim. *2*, 322 (1973)
4.39 G.B. Sergeyev, V.A. Batyuk, M.B. Stepanov, B.M. Sergeyev: Dokl. Akad. Nauk SSSR *213*, 891 (1973)
4.40 A.Yu. Shaulov, A.B. Shapiro, A.G. Sklarova, A.M. Wasserman, A.L. Buchachenko, E.G. Rozantsev: Vysokomolek. Soyed. *16A*, 2680 (1974)
4.41 G.M. Zhidomirov, A.L. Buchachenko: Zh. Strukt. Khim. *8*, 1110 (1967)
V.N. Parmon, A.I. Kokorin, G.M. Zhidomirov: Zh. Strukt. Khim. *18*, 132 (1977)
4.42 E.G. Rozantsev: *Free Iminoxyl Radicals* (Khimiya, Moscow 1970)
4.43 V.N. Parmon, A.I. Kokorin, G.M. Zhidomirov, K.I. Zamaraev: Mol. Phys. *30*, 659 (1975)
4.44 A.I. Kokorin, V.N. Parmon: Izv. Akad. Nauk SSSR, Ser. Khim. *3*, 576 (1978)
4.45 G.I. Skubnevskaya, K.M. Salikhov, L.M. Smirnova, Yu.N. Molin: Kinet. Katal. *11*, 888 (1970)
4.46 G.N. La Mar, W.D. Horrocks, Jr., R.H. Holm: *NMR of Paramagnetic Molecules - Principles and Applications* (Academic Press, New York 1973)
4.47 Z. Luz, B.L. Silver, D. Fiat: J. Chem. Phys. *46*, 469 (1967)
4.48 D. Fiat, Z. Luz, B.L. Silver: J. Chem. Phys. *49*, 1376 (1968)
4.49 A. Obynochny: Candidate of Sciences Thesis, Novosibirsk (1976)
4.50 P.W. Anderson: Phys. Rev. *115*, 2 (1959)
4.51 J. Owen: J. Appl. Phys. *32*, 213 (1961)
4.52 R.P. Shibayeva: Zh. Strukt. Khim. *16*, 330 (1975)
4.53 A.I. Kokorin, V.N. Parmon, V.I. Suskina, Yu.A. Ivanov, E.G. Rozantsev, K.I. Zamaraev: Zh. Fiz. Khim. *48*, 953 (1974)
4.54 A.G. Pearson, T. Buch: J. Chem. Phys. *36*, 1277 (1962)
4.55 M.T. Jones: J. Chem. Phys. *38*, 2892 (1963)
4.56 K.I. Zamaraev, A.T. Nikitaev: Kinet. Katal. *11*, 797 (1970)
4.57 K.I. Zamaraev, A.T. Nikitaev, G.A. Senyukova: Kinet. Katal. *13*, 54 (1972)
4.58 K.I. Zamaraev: Doctor of Sciences Thesis, Moscow (1972)
4.59 J.G. Powles, M.H. Mosley: Proc. Phys. Soc. *78*, 370 (1961)
4.60 J.C. Danner, T.R. Tuttle, Jr.: J. Am. Chem. Soc. *85*, 4052 (1963)
4.61 M.T. Watts, M.L. Lu, M.P. Eastman: J. Phys. Chem. *77*, 625 (1973)
4.62 M.A. Komarinsky, A.C. Wahl: J. Phys. Chem. *79*, 695 (1975)
4.63 S.A. Al'tshuler, B.M. Kozyrev: *Electron Paramagnetic Resonance of Compounds of Intermediate Group Elements* (Nauka, Moscow 1972)
4.64 A.K. Pikayev: *Solvated Electron in Radiation Chemistry* (Nauka, Moscow 1969)

4.65 M.T. Watts, M.L. Lu, R.C. Chen, M.P. Eastman: J. Phys. Chem. *77*, 2959 (1973)
4.66 P. Debye: Trans. Electrochem. Soc. *82*, 265 (1942)
4.67 A.T. Nikitaev: Candidate of Sciences Thesis, Moscow (1972)
4.68 G.I. Likhtenstein: *The Method of Spin Labels in Molecular Biology* (Nauka, Moscow 1974)
4.69 R.E. Barnett, C.M. Grisham: Bioch. Bioph. Res. Comm. *48*, 1362 (1972)
4.70 P. Devaux, H.M. McConnell: J. Am. Chem. Soc. *94*, 4475 (1972)
4.71 E. Sackmann, H. Träuble: J. Am. Chem. Soc. *94*, 4482 (1972)
4.72 P. Devaux, C.J. Scandella, H.M. McConnell: J. Mag. Reson. *9*, 474 (1973)
4.73 V.I. Leventahl, J.M. Backer, Yu.N. Molin, V.P. Kumarev, M.A. Grachev, D.G. Knorre: FEBS Lett. *24*, 149 (1972)
4.74 Yu.B. Grebenshchikov, G.V. Ponomarev, R.P. Yevstigneyeva, G.I. Likhtenshtein: Biofizika *17*, 910 (1972)
4.75 L.M. Reichman, B. Annayev, E.G. Rozantsev: Bioch. Bioph. Acta *263*, 41 (1972)
4.76 H.H. Ruf, W. Nastainczyk, V. Ulrich: 6th Int. Conf. on Magn. Resonance in Biol. Systems, Abst. C. 14, Kandersteg (1974)
4.77 H.H. Ruf, W. Nastainczyk: Eur. J. Biochim. *66*, 139 (1976)
4.78 A.D. Keith, W. Snipes, D. Chapman: Biochem. *16*, 634 (1977)
4.79 A.D. Keith, W. Snipes: Science *183*, 666 (1974)
4.80 P. Morse, II: Bioch. Bioph. Res. Comm. *77*, 1486 (1977)
4.81 P.D. Morse, II, M. Ruhling, W. Snipes, A.D. Keith: Arch. Bioch. Bioph. *168*, 40 (1975)
4.82 J.M. Backer, V.G. Budker, S.I. Yeremenko, Yu.N. Molin: Dokl. Akad. Nauk SSSR *230*, 466 (1976)
4.83 Yu.E. Kirsh, V.Ya. Kovner, A.I. Kokorin, K.I. Zamaraev, V.A. Kabanov: Dokl. Akad. Nauk SSSR *212*, 138 (1973)
4.84 G.K. Humphries, H.M. McConnell: Proc. Nat. Acad. Sci. *71*, No 5, 1691 (1974)
4.85 V.V. Lyachovich, I.B. Tsyrlov, V.V. Mishin, L.M. Weiner, G.V. Rumjantzeva, V.G. Budker: Acta Medic. Germ. *38*, 201 (1979)

Subject Index

MIX
Papier aus verantwortungsvollen Quellen
Paper from responsible sources
FSC® C105338

If you have any concerns about our products,
you can contact us on
ProductSafety@springernature.com

In case Publisher is established outside the EU,
the EU authorized representative is:
Springer Nature Customer Service Center GmbH
Europaplatz 3, 69115 Heidelberg, Germany

Printed by Libri Plureos GmbH
in Hamburg, Germany